Lecture Notes in Mecha

Editorial Board Member

Francisco Cavas-Martínez, Departamento de Estructuras, Construcción y Expresión Gráfica Universidad Politécnica de Cartagena, Cartagena, Murcia, Spain

Series Editor

Fakher Chaari, National School of Engineers, University of Sfax, Sfax, Tunisia

Editorial Board Member

Francesca di Mare, Institute of Energy Technology, Ruhr-Universität Bochum, Bochum, Nordrhein-Westfalen, Germany

Series Editor

Francesco Gherardini, Dipartimento di Ingegneria "Enzo Ferrari", Università di Modena e Reggio Emilia, Modena, Italy

Editorial Board Member

Mohamed Haddar, National School of Engineers of Sfax (ENIS), Sfax, Tunisia

Series Editor

Vitalii Ivanov, Department of Manufacturing Engineering, Machines and Tools, Sumy State University, Sumy, Ukraine

Editorial Board Members

Young W. Kwon, Department of Manufacturing Engineering and Aerospace Engineering, Graduate School of Engineering and Applied Science, Monterey, CA, USA

Justyna Trojanowska, Poznan University of Technology, Poznan, Poland

Lecture Notes in Mechanical Engineering (LNME) publishes the latest developments in Mechanical Engineering—quickly, informally and with high quality. Original research reported in proceedings and post-proceedings represents the core of LNME. Volumes published in LNME embrace all aspects, subfields and new challenges of mechanical engineering. Topics in the series include:

- Engineering Design
- Machinery and Machine Elements
- Mechanical Structures and Stress Analysis
- Automotive Engineering
- Engine Technology
- Aerospace Technology and Astronautics
- Nanotechnology and Microengineering
- Control, Robotics, Mechatronics
- MEMS
- Theoretical and Applied Mechanics
- Dynamical Systems, Control
- Fluid Mechanics
- Engineering Thermodynamics, Heat and Mass Transfer
- Manufacturing
- Precision Engineering, Instrumentation, Measurement
- Materials Engineering
- Tribology and Surface Technology

To submit a proposal or request further information, please contact the Springer Editor of your location:

China: Ms. Ella Zhang at ella.zhang@springer.com
India: Priya Vyas at priya.vyas@springer.com
Rest of Asia, Australia, New Zealand: Swati Meherishi at swati.meherishi@springer.com
All other countries: Dr. Leontina Di Cecco at Leontina.dicecco@springer.com

To submit a proposal for a monograph, please check our Springer Tracts in Mechanical Engineering at https://link.springer.com/bookseries/11693 or contact Leontina.dicecco@springer.com

Indexed by SCOPUS. All books published in the series are submitted for consideration in Web of Science.

More information about this series at https://link.springer.com/bookseries/11236

Magdalena Diering · Michał Wieczorowski ·
Mukund Harugade · Alejandro Pereira
Editors

Advances in Manufacturing III

Volume 4 - Measurement and Control Systems: Research and Technology Innovations, Industry 4.0

Springer

Editors
Magdalena Diering
Faculty of Mechanical Engineering
Poznan University of Technology
Poznan, Poland

Mukund Harugade
Department of Mechanical Engineering
Birla Institute of Technology and Science
Pilani, India

Michał Wieczorowski
Faculty of Mechanical Engineering
Poznan University of Technology
Poznan, Poland

Alejandro Pereira
Faculty of Industrial Engineering
University of Vigo
Zamans, Pontevedra, Spain

ISSN 2195-4356 ISSN 2195-4364 (electronic)
Lecture Notes in Mechanical Engineering
ISBN 978-3-031-03924-9 ISBN 978-3-031-03925-6 (eBook)
https://doi.org/10.1007/978-3-031-03925-6

© The Editor(s) (if applicable) and The Author(s), under exclusive license to Springer Nature Switzerland AG 2022

This work is subject to copyright. All rights are solely and exclusively licensed by the Publisher, whether the whole or part of the material is concerned, specifically the rights of translation, reprinting, reuse of illustrations, recitation, broadcasting, reproduction on microfilms or in any other physical way, and transmission or information storage and retrieval, electronic adaptation, computer software, or by similar or dissimilar methodology now known or hereafter developed.

The use of general descriptive names, registered names, trademarks, service marks, etc. in this publication does not imply, even in the absence of a specific statement, that such names are exempt from the relevant protective laws and regulations and therefore free for general use.

The publisher, the authors and the editors are safe to assume that the advice and information in this book are believed to be true and accurate at the date of publication. Neither the publisher nor the authors or the editors give a warranty, expressed or implied, with respect to the material contained herein or for any errors or omissions that may have been made. The publisher remains neutral with regard to jurisdictional claims in published maps and institutional affiliations.

This Springer imprint is published by the registered company Springer Nature Switzerland AG
The registered company address is: Gewerbestrasse 11, 6330 Cham, Switzerland

Preface

This volume of Lecture Notes in Mechanical Engineering contains selected papers presented at the 7th International Scientific-Technical Conference MANUFACTURING 2022, held in Poznan, Poland, on May 16–19, 2022. The conference was organized by the Faculty of Mechanical Engineering, Poznan University of Technology, Poland.

The aim of the conference was to present the latest achievements in mechanical engineering and to provide an occasion for discussion and exchange of views and opinions. The main conference topics are as follows:

- mechanical engineering
- production engineering
- quality engineering
- measurement and control systems
- biomedical engineering.

The organizers received 165 contributions from 23 countries around the world. After a thorough peer-review process, the committee accepted 91 papers for conference proceedings prepared by 264 authors from 23 countries (acceptance rate around 55%). Extended versions of selected best papers will be published in the following journals: *Management and Production Engineering Review, Bulletin of the Polish Academy of Sciences: Technical Sciences, Materials, Applied Sciences.*

The book **Advances in Manufacturing III** is organized into five volumes that correspond with the main conference topics mentioned above.

Advances in Manufacturing III - Volume 4 - Measurement and Control Systems: Research and Technology Innovations, Industry 4.0 gathers extensive information for both academics and practitioners on methods and tools for metrology, measurement and control, and their applications in many fields such as biology, chemistry and food science, material science, and manufacturing and civil engineering. It also covers advanced computer science methods for measurement, control, and quality assurance. Contributions were reviewed by experts from various areas to ensure a selection of high-quality content, with relevance for both academics and professionals. This book includes 14 chapters, prepared by 52 authors from seven countries.

We would like to thank the members of the international program committee for their hard work during the review process.

We acknowledge all that contributed to the staging of MANUFACTURING 2022: authors, committees, and partners. Their involvement and hard work were crucial to the success of the MANUFACTURING 2022 conference.

May 2022

Magdalena Diering
Michał Wieczorowski
Mukund Harugade
Alejandro Pereira

Organization

Steering Committee

General Chair

Adam Hamrol — Poznan University of Technology, Poland

Chairs

Olaf Ciszak — Poznan University of Technology, Poland
Magdalena Diering — Poznan University of Technology, Poland
Justyna Trojanowska — Poznan University of Technology, Poland

Scientific Committee

Ahmad Majdi Abdul-Rani — University of Technology PETRONAS, Malaysia
Dario Antonelli — Politecnico di Torino, Italy
Katarzyna Antosz — Rzeszow University of Technology, Poland
Jorge Bacca-Acosta — Fundación Universitaria Konrad Lorenz, Colombia
Zbigniew Banaszak — Koszalin University of Technology, Polska
Andre D. L. Batako — Liverpool John Moores University, UK
Stefan Berczyński — West Pomeranian University of Technology in Szczecin, Poland
Alain Bernard — Ecole Centrale de Nantes, France
Kristina Berladir — Sumy State University, Ukraine
Christopher Brown — Worcester Polytechnic Institute, USA
Anna Burduk — Wrocław University of Science and Technology, Poland
Marcin Butlewski — Poznan University of Technology, Poland
Lenka Čepová — VSB Technical University of Ostrava, Czech Republic
Nadežda Čuboňová — University of Žilina, Slovakia

Somnath Chattopadhyaya	Indian School of Mines, India
Olaf Ciszak	Poznań University of Technology, Polska
Jacek Diakun	Poznań University of Technology, Poland
Magdalena Diering	Poznan University of Technology, Poland
Ewa Dostatni	Poznan University of Technology, Poland
Jan Duda	Cracow University of Technology, Poland
Davor Dujak	J.J. Strossmayer University of Osijek, Croatia
Milan Edl	University of West Bohemia, Czech Republic
Brigita Gajšek	University of Maribor, Slovenia
Mosè Gallo	University of Naples Federico II, Italy
Adam Gałuszka	Silesian University of Technology, Poland
Józef Gawlik	Cracow University of Technology, Poland
Adam Gąska	Cracow University of Technology, Poland
Adam Górny	Poznan University of Technology, Polska
Filip Górski	Poznan University of Technology, Poland
Adam Hamrol	Poznan University of Technology, Poland
Mukund Harugade	Padmabhooshan Vasantraodada Patil Institute of Technology, India
Ivan Hudec	Slovak University of Technology in Bratislava, Slovakia
Jozef Husár	Technical University of Košice, Slovakia
Aminul Islam	Technical University of Denmark, Denmark
Vitalii Ivanov	Sumy State University, Ukraine
Andrzej Jardzioch	West Pomeranian University of Technology in Szczecin, Poland
Sabahudin Jašarević	University of Zenica, Bosnia and Herzegovina
Małgorzata Jasiulewicz-Kaczmarek	Politechnika Poznańska, Poland
Wojciech Kacalak	Technical University of Koszalin, Poland
Sławomir Kłos	University of Zielona Góra, Polska
Lucia Knapčíková	Technical University of Košice, Slovakia
Damian Krenczyk	Silesian University of Technology, Poland
Jolanta Krystek	Silesian University of Technology, Poland
Józef Kuczmaszewski	Lublin University of Technology, Poland
Agnieszka Kujawińska	Poznan University of Technology, Poland
Ivan Kuric	University of Žilina, Slovak Republic, Slovak Republic
Piotr Łebkowski	AGH, University of Science and Technology, Poland
Stanisław Legutko	Poznan University of Technology, Polska
Sławomir Luściński	Kielce University of Technology, Poland
José Machado	University of Minho, Portugal
Vijaya Kumar Manupati	National Institute of Technology, Warangal, India
Uthayakumar Marimuthu	Kalasalingam Academy of Research and Education, India

Jorge Martín Gutiérrez	Universidad de La Laguna, Spain
Józef Matuszek	University of Bielsko-Biala, Poland
Arkadiusz Mężyk	Silesian University of Technology, Poland
Dariusz Mikołajewski	Kazimierz Wielki University, Poland
Andrzej Milecki	Poznań University of Technology, Poland
Piotr Moncarz	Stanford University, USA
Manurl Francisco Morales-Contreras	Universidad Pontificia Comillas, Spain
Keddam Mourad	USTHB, Algeria
Leticia Neira	UANL, México
Przemysław Niewiadomski	University of Zielona Góra, Polska
Erika Ottaviano	University of Cassino and Southern Lazio, Cassino
Eren Özceylan	Gaziantep University, Turkey
Razvan Ioan Pacurar	Technical University of Cluj-Napoca, Romania
Justyna Patalas-Maliszewska	University of Zielona Góra, Poland
Ivan Pavlenko	Sumy State University, Ukraine
Lucjan Pawłowski	Lublin University of Technology; Polish Academy of Science, Poland
Dragan Peraković	University of Zagreb, Croatia
Alejandro Pereira	Universidade de Vigo, Spain
Marko Periša	University of Zagreb, Croatia
Pierluigi Rea	University of Cagliari, Italy
Ján Piteľ	Technical University of Košice, Slovakia
Izabela Rojek	Kazimierz Wielki University, Poland
Alessandro Ruggiero	University of Salerno, Italy
Dominik Rybarczyk	Poznań University of Technology, Polska
Krzysztof Santarek	Warsaw University of Technology, Poland
Jarosław Sęp	Rzeszow University of technology, Poland
Bożena Skołud	Silesian University of Technology, Polska
Jerzy Andrzej Sładek	Cracow University of Technology, Poland
Dorota Stadnicka	Rzeszów University of Technology, Poland
Beata Starzyńska	Poznan University of Technology, Poland
Manuel Francisco Suárez Barraza	Universidad de las Américas Puebla, Mexico
Marcin Suszyński	Poznan University of Technology, Polska
Marek Szostak	Poznan University of Technology, Poland
Erfan Babaee Tirkolaee	Istinye University, Turkey
Justyna Trojanowska	Poznan University of Technology, Poland
Leonilde Varela	University of Minho, Portugal
Katarzyna Węgrzyn-Wolska	Engineering School of Digital Technologies, France
Gerhard-Wilhelm Weber	Poznan University of Technology, Poland
Edmund Weiss	Calisia University, Poland
Dorota Więcek	University of Bielsko-Biala, Poland

Michał Wieczorowski	Poznan University of Technology, Poland
Hanna Włodarkiewicz-Klimek	Poznan University of Technology, Poland
Szymon Wojciechowski	Poznan University of Technology, Poland
Ralf Woll	BTU Cottbus-Senftenberg, Germany
Jozef Zajac	Košice University of Technology, Slovakia
Nermina Zaimovic-Uzunovic	University of Zenica, Bosnia, and Herzegovina

Program Committee

Karla Alvarado-Ramírez
José Angel
Dario Antonelli
Katarzyna Antosz
Erfan Babaee Tirkolaee
Jorge Bacca-Acosta
Diana Baila
Prashanth Bandari
Lucía Barcos-Redín
Andre Batako
Petr Beneš
Kristina Berladir
Marcin Białek
Mikołaj Bilski
Matej Borovinšek
Anna Borucka
Cristina Borzan
Sara Bragança
Daniel Brissaud
Christopher Brown
Paweł Buń
Anna Burduk
Bartłomiej Burlaga
Jacek Buśkiewicz
Marcin Butlewski
Massimiliana Carello
Fernando Castillo
Robert Cep
Miroslav Cisar
Olaf Ciszak
Felipe Contreras
Tomáš Coranič
Eric Costa
Margareta Coteata

Dorota Czarnecka-Komorowska
Reggie Davidrajuh
Mario del Valle
Boris Delibašić
Jakub Demčák
Jacek Diakun
Magdalena Diering
Grzegorz Domek
Ewa Dostatni
Radosław Drelich
Jan Duda
Luboslav Dulina
Eduard Franas
Paweł Fritzkowski
Brigita Gajšek
Mosè Gallo
Bartosz Gapiński
Katarzyna Gawdzińska
Adam Gąska
Dominik Gojdan
Arkadiusz Gola
Fernando González-Aleu
Adam Górny
Filip Górski
Cezary Grabowik
Marta Grabowska
Jakub Grabski
Damian Grajewski
Patrik Grznár
Aleksander Gwiazda
Michal Hajžman
Adam Hamrol
Mukund Harugade
Amalija Horvatić-Novak

Stella Hrehová
Jozef Husár
Vitalii Ivanow
Carmen Jaca
Rajat Jain
Michał Jakubowicz
Małgorzata Jankowska
Andrzej Jardzioch
Małgorzata Jasiulewicz-Kaczmarek
Anna Karwasz
Jakub Kaščak
Sławomir Kłos
Lucia Knapčíková
Adam Koliński
Elif Kongar
Boris Kostov
Kateryna Kovbasiuk
Arkadiusz Kowalski
Tomasz Kowaluk
Damian Krenczyk
Grzegorz Królczyk
Józef Kuczmaszewski
Agnieszka Kujawińska
Panagiotis Kyratsis
Georgios Lampropoulos
Peter Lazorík
Stanislaw Legutko
Andrzej Loska
Czesław Łukianowicz
José Machado
Marek Macko
Ján Majerník
Vitalii Maksymenko
Damjan Maletič
Matjaž Maletič
Vijaya Kumar Manupati
Sven Maricic
Jorge Martín Gutiérrez
Thomas Mathia
Maciej Matuszewski
Dariusz Mazurkiewicz
Jakub Michalski
Dariusz Mikołajewski
Janusz Mleczko
Ladislav Morovic

Keddam Mourad
Leticia Neira
Zdeněk Neusser
Magdalena Niemczewska-Wójcik
Przemysław Niewiadomski
Czesław Niżankowski
Nejc Novak
Filip Osiński
Erika Ottaviano
Eren Ozceylan
Ancuta Pacurar
Kyratsis Panagiotis
Jason Papathanasiou
Waldemar Paszkowski
Justyna Patalas-Maliszewska
Ivan Pavlenko
Paweł Pawlus
Jarosław Pempera
Dragan Peraković
Alejandro Pereira
Dariusz Plinta
Florin Popister
Evangelos Psomas
Paulina Rewers
Francisco Gabriel Rodriguez-González
Michał Rogalewicz
Izabela Rojek
Mirosław Rucki
Biserka Runje
Dominik Rybarczyk
Michał Rychlik
Milan Saga
Alžbeta Sapietova
Filip Sarbinowski
Holger Schlegel
Armin Schleinitz
Dariusz Sędziak
Robert Sika
Zbyněk Šika
Varinder Singh
Bożena Skołud
António Lucas Soares
Ewa Stachowska
Dorota Stadnicka
Sergiu Dan Stan

Roman Starosta
Beata Starzyńska
Krzysztof Stępień
Tomasz Stręk
Manuel Suárez-Barraza
Marcin Suszynski
Grażyna Sypniewska-Kamińska
Marek Szostak
Jiří Tengler
Samala Thirupathi
Francisco Torres
Justyna Trojanowska
Piotr Trojanowski
Krzysztof Tyszczuk
M. Uthayakumar
Leonilde Varela
Nikola Vitkovic

Adrián Vodilka
Sachin Waigaonkar
Tomasz Walczak
Łukasz Warguła
Gerhard-Wilhelm Weber
Katarzyna Węgrzyn-Wolska
Radosław Wichniarek
Michał Wieczorowski
Dariusz Więcek
Hanna Włodarkiewicz-Klimek
Szymon Wojciechowski
Adam Woźniak
Ryszard Wyczółkowski
Nermina Zaimovic-Uzunovic
Ivan Zajačko
Magdalena Żukowska
Krzysztof Żywicki

Special Sessions

Manufacturing and Management Approaches and Tools for Companies in Era of Digitalization

Special Session Organizing Committee

Leonilde Varela	University of Minho, Portugal
Justyna Trojanowska	Poznan University of Technology, Poland
Vijaya Kumar Manupati	Mechanical Engineering Department, India
Paulina Rewers	Poznan University of Technology, Poland

Operational Innovation and Excellence in Organizations of Twenty-First Century. Lean Thinking, Kaizen Philosophy, and Toyota Production System

Manuel F. Suárez-Barraza	Universidad de las Américas Puebla, Mexico
Manuel F. Morales-Contreras	Universidad Pontificia Comillas, Spain
Agnieszka Kujawińska	Poznan University of Technology, Poland
Marta Grabowska	Poznan University of Technology, Poland

Cyber-Physical Production Systems: From Design to Applications

Katarzyna Antosz	Rzeszow University of Technology, Poland
José Machado	University of Minho, Portugal
Erika Ottaviano	University of Cassino and Southern Lazio, Italy
Pierluigi Rea	University of Cagliari, Italy

Augmented, Virtual and Mixed Reality Systems

Jozef Husár	Technical University of Košice, Slovak Republic
Jakub Kaščak	Technical University of Košice, Slovak Republic
Paweł Buń	Poznan University of Technology, Poland

Intelligent Methods Supporting Manufacturing Systems Efficiency

Anna Burduk	Wroclaw University of Science and Technology, Poland
Andre Batako	Liverpool John Moores University, UK
Dorota więcek	University of Bielsko-Biala, Poland
Ivan Kuric	University of Žilina, Slovak Republic

Sustainability on Production in the Aspect of Industry 4.0

Ewa Dostatni	Poznan University of Technology, Poland
Izabela Rojek	Kazimierz Wielki University in Bydgoszcz, Poland
Jacek Diakun	Poznan University of Technology, Poland
Dariusz Mikołajewski	Kazimierz Wielki University in Bydgoszcz, Poland

Design and Rapid Manufacturing of Customized Medical Products

Filip Górski	Poznan University of Technology, Poland
Magdalena Żukowska	Poznan University of Technology, Poland
Razvan Pacurar	Technical University of Cluj-Napoca, Romania

Metrology and Statistical Analysis of Measurement and Control Systems

Alejandro Pereira Domínguez	University of Vigo, Spain
Lenka Čepová	VSB—Technical University of Ostrava, Czech Republic
Magdalena Diering	Poznan University of Technology, Poland
Bartosz Gapiński	Poznan University of Technology, Poland

Analytical and Computational Methods in (Bio-) Mechanical and Material Engineering

Tomasz Stręk	Poznan University of Technology, Poland
Roman Starosta	Poznan University of Technology, Poland
Nejc Novak	University of Maribor, Slovenia
Pavel Polach	University of West Bohemia, Czech Republic

Contents

Impact of the Type of Tool Holder on the Surface Finish During Precision Machining of WCLV Hardened Steel with Carbide Milling Cutters 1
Joanna Krajewska-Śpiewak, Józef Gawlik, Marcin Grabowski, Piotr Tyczyński, and Małgorzata Kowalczyk

Analysis and Modelling of Intelligent Gate Valve with Fault Detection and Prevention System for Fabrication Using 3D Printing Method 10
Bushra Mohamad Zakir, Ahmad Majdi A. Rani, Nu'man Al-Basyir Mohd Nuri Al-Amin, Mohd Norhisyam Awang, M. Nasir A. Rani, Nur Athirah Ishar, Harvey M. Thompson, Marek Szostak, and Krishnan Subrmaniam

Effect of Combined Mechanical and Electrochemical Action on Surface Roughness in Microturning Process 26
Marcin Grabowski, Sebastian Skoczypiec, and Dominik Wyszynski

Evaluation of the Fidelity of Additively Manufactured 3D Models of a Fossil Skull 36
Miroslaw Rucki, Yaroslav Garashchenko, Ilja Kogan, and Tomasz Ryba

Analysis of the Direction-Dependent Point Identification Accuracy in CMM Measurement 48
Tomasz Mazur, Miroslaw Rucki, Michal Jakubowicz, and Lenka Cepova

Non-destructive Testing of Metal-Polymer Laminates by Digital Shearography 57
Zuzanna Konieczna, Frans Meijer, and Ewa Stachowska

Performance of Selected Thermally Sprayed Coatings for Power Applications 67
Šárka Houdková, Zdeněk Česánek, and Pavel Polach

Laser Measurement by Angle Accuracy Method in Additive Technology SLM 316L 79
Ondrej Mizera, Jiri Hajnys, Lenka Cepova, Jan Zelinka,
and Jakub Mesicek

Review of Measurement Methods to Evaluate the Geometry of Different Types of External Threads 89
Bartłomiej Krawczyk, Krzysztof Smak, Piotr Szablewski,
and Bartosz Gapiński

New Methodology of Face Mill Path Correction Based on Machined Surface Measurement to Improve Flatness..................... 102
Marek Rybicki

Development of the Handheld Measuring Probe for a 3D Scanner 115
Robert Kupiec, Wiktor Harmatys, Izabela Sanetra,
Katarzyna Składanowska, and Ksenia Ostrowska

Comparison of Measurements Realized on Computed Tomograph and Optical Scanners for Elements Manufactured by Wire Arc Additive Manufacturing 127
Michał Wieczorowski, I. P. Yago, Pereira Domínguez Alejandro,
Bartosz Gapiński, Grzegorz Budzik, and Magdalena Diering

Verification of Computed Tomograph for Dimensional Measurements 142
Bartosz Gapiński, Michał Wieczorowski, Patryk Mietliński,
and Thomas G. Mathia

Identification of the Sensitivity of FDM Technology to Material Moisture with a Fast Test 156
Adam Hamrol, Maciej Cugier, and Filip Osiński

Author Index... 167

Impact of the Type of Tool Holder on the Surface Finish During Precision Machining of WCLV Hardened Steel with Carbide Milling Cutters

Joanna Krajewska-Śpiewak[1(✉)], Józef Gawlik[1], Marcin Grabowski[1], Piotr Tyczyński[2], and Małgorzata Kowalczyk[1]

[1] Cracow University of Technology, Kraków, Poland
joannakrajewska.pk@gmail.com, malgorzata.kowalczyk@pk.edu.pl
[2] Limatherm S.A, ul. Tarnowska 1, 34-600 Limanowa, Poland
piotr.tyczynski@mapal.com

Abstract. Precision milling of curvilinear free surfaces with the use of monolithic milling cutters is used, among others, in dies and molds production and in production of press-forming dies made of hardened steel. Precision milling processes are carried out with the following milling parameters: axial depth of cut a_p < 0.33 mm and radial depth of cut a_e < 0.5 mm, with the required machining accuracy below 40 µm. The surface quality of injection molds has a direct impact on the quality of the obtained element (moulded piece). Thanks to the use of carbide tools, high reliability and machining quality are obtained, which allows to eliminate the grinding process. The tests were carried out on samples made of WCLV steel with a hardness of 45–47 HRC (in the Rockwell C scale) with double-edge end mill with a diameter of 6 mm. Two types of tool holders were used to mount a milling cutter. Heat-shrinkable holders and holders made with the use of incremental laser sintering technology were used. Recommendations for the selection of tool holders and machining conditions were defined, which allowed to obtain the required technological quality of the machining surfaces.

Keywords: Tool holder · Precision machining · Hardened tool steel

1 Introduction

A number of factors decide about the beginning of the process of disrupting the consistency of the material that determines the minimal thickness of the cutting layer. The main ones are the properties of the workpiece, the cutting-edge geometry and the rigidity of the machining system.

Hardened WCLV steel with a hardness of about 50 HRC belongs to the group of difficult-to-cut materials. In the process of machining, of such materials, an important role is played by the certainty of the milling cutter fixture in the tool holder.

Carbide end mills with a diameter of 6 mm were used in the tests. Two types of tool holders (collets) were used to mount the milling cutter: sintered tool holders made with the incremental technology and heat-shrinkable tool holders.

Hot work tool WCLV steel is used on dies for precise, high-pressure casting of aluminum alloys thin-walled products. The presented research results, for the conducted analysis of the literature, relate to the machining of difficult-to-cut materials.

The process of separating the cutting material into a chip is dynamic. The authors [1] assessed the results of experimental studies on determining the minimal thickness of the cutting layer during machining with single-edge tools with defined stereometry. They developed dependencies for determining the minimal thickness of the cutting layer under the condition of constant temperature. They pointed out that the measurement results are influenced by a number of disturbing factors, which depend on both the properties of the workpiece and of the tool.

The authors [2] present the results of physical tests and finite element method (FEM) simulation of Inconel 718 face milling. A significant influence of feed and cutting speed on the wear of a carbide insert with a TiAlN/AlCrN coating was demonstrated. It was also stated that the type of formed chip depends on the cutting conditions. Segmented sawtooth chip or discontinuous toothed (stepped) chip strongly influences the wear pattern of the cutting edges.

In a paper [3], the authors emphasize that precise, finishing processing of hardened surfaces requires the use of tools with sintered carbide cutting edges. The occurrence of the decohesion process is decisively influenced by: the radius of the rounding of the cutting edge and the cross-section of the cutting layer. The cutting-edge radius and the cross-section of the cutting layer depend on the cutting parameters and influence the occurrence of the decohesion process. The Analysis of Variance (ANOVA) Taguchi methodology was used in the research program. Similarly, in the article [4] the authors used the Taguchi experiment planning method to investigate the surface quality in the process of CNC face milling.

In the experimental studies of the micro-milling process of nickel super alloy 718, the authors [5] showed that the abrasive wear of the side face of the cutter is dominant. The fatigue wear of cutters also occurred. It is possible to reduce the wear of the edges by applying appropriate coatings on the working surfaces of the cutters.

The authors of the publication [6] presented the original method of identifying the minimal thickness of the cutting layer with the use of the acoustic emission signal under the conditions of machining with multi-edges milling cutters for Inconel 718 and Ti6Al4V alloys. The usefulness of this method has been demonstrated in a real production conditions. The results of these tests complete and extend the methods of the minimal thickness of the cutting layer identification [7, 8].

The authors [9] analyzed and assessed the influence of cutting parameters on the wear topography and the efficiency of the Inconel 718 alloy machining process with the use of various types of carbide inserts. The results of these studies can be used in the multi-criteria optimization of the selection of machining conditions based on the criteria for the durability of the cutting edges and the efficiency of the process.

Machining of hardened steels was the main subject of the research carried out by the authors of the publications [10, 11]. The test results confirm that machining of hardened materials, with a hardness of up to 50 HRC, with tools of defined tool stereometry can replace grinding operations. The authors specified some of the most important benefits, regarding this solution, such as increased productivity, flexibility, lower capital expenditure and reduction of environmentally harmful waste.

The authors of the article [12] present the results of tests of tools with uncoated carbide edges and tools with cutting edges with a multi-layer chemical vapor deposition coating (CVD) during dry machining of the Ti-6242S titanium alloy which is used, among others, in aircraft engines. The topography and the wear mechanism of the edges were determined. The finite element method was used to analyze the tribological parameters and chip segmentation. It was the basis for proposing the mechanisms of coating delamination.

The authors [13] presented the results of experimental and numerical research of titanium-based alloy (Ti-6Al-4V) subjected to face milling. The predicted course of the cutting-edge wear, based on the 3D simulation with the finite element method, shows good compliance with the experimental measurements of the cutting edge wear. It was also found that the residual stresses in the surface layer of the machining surface have compressive character.

In the publication [14] the authors present the results of the study of the influence of cutting conditions and cutting-edge wear on the surface condition of titanium alloys after dry milling. The tests were carried out on tools with a TiAlN + TiN coating made by the physical vapor deposition method (PVD). Samples were subjected to the following values of cutting speeds $v_c = (100; 300)$ m/min and feeds $v_f = (0.03; 0.06)$ mm/tooth. The use of higher cutting speed and feed as well as the cutting-edge wear caused an increase in the microhardness of the surface layer of the workpiece.

The authors [15] conducted tests of CNC milling of the Inconel 718 alloy with solid carbide milling cutters with TiAlN (PVD) coating under cryogenic cooling conditions. The roughness of the machined surface was reduced from 33% to 40%. At the same time, cryogenic cooling significantly reduced the life of carbide tools due to the occurrence of chipping and cracks in the cutting edges.

An extensive analysis of the literature in the field of milling, quality assessment of machined surfaces and used tools can be found in the publication [16]. The authors emphasize that the roughness of the machined surfaces has a significant impact on the wear resistance and fatigue strength of the obtained elements.

The test results confirm that mounting the cutters in the tool holder affects their work. The authors of the publication [17, 18] showed that changing the stiffness of spherical cutters by changing the extension length from the holder has a significant impact on the displacement of the working part of the cutter. This directly affects the values of the vibration amplitudes, and thus the roughness parameters of the machined surface.

The authors of the publication [19] showed that high-speed machining (HSM) of tool steels in the hardened state is an attractive solution for the mold and die industry. More than 50% of the total production time is attributed to traditional machining process and polishing. Significant cost savings and productivity improvements can be achieved with the use of HSM.

The conducted analysis of the literature, on precision machining of difficult-to-cut materials, allowed the authors to notice the lack of information on the operational characteristics of the tool holder used for mounting end mills for machining process of hardened WCLV steel. Therefore, it was justified to undertake research that results from the needs of the industry.

2 Experimental Details

The cutting tool wear in precision machining is one of the key factors which affects the quality of the obtained details and the stability of the machining process. The emerging cracks or chipping of the cutting edge during the cutting process are difficult to detect and can cause deterioration of the surface quality. This may lead to the recognition of the detail as a broken component. In order to meet the high-quality requirements, set for injection molds, it is necessary to determine the impact of the cutting-edge wear on the condition of the workpiece surface.

Tests were carried out on samples made of WCLV steel (Table 1) with a hardness of 45–47 HRC and dimensions of 121 × 121 × 11 mm with an end mill with a diameter of ϕ6 mm. The geometry of the carbide milling cutter was selected on the basis of previous machine tests. A solid, right-handed carbide milling cutter with a rake angle of 3° was used. Based on the preliminary experiments [20], the cutting tests were carried out under the finishing conditions for the following machining parameters:
$a_p = 0.04$ mm, $a_e = 0.04$ mm, $v_f = 1200 \frac{mm}{min}$, $n = 9000$ rpm.

Table 1. WCLV steel chemical composition.

C	Si	Mn	P	S	Cr	Mo	V
0.35–0.42	0.80–1.20	0.25–0.50	max 0.03	max 0.03	4.80–5.50	1.20–1.50	0.85–1.15

The cutting-edge wear was determined on the basis of the obtained 3D scans of the tool before and after machining. Measurement of the tool wear after a certain cutting time, taking scans and reassembling the tools, could lead to some measurement errors. In order to avoid these errors, it was decided to measure tools that had been working for a certain period of time in order to determine the tool wear. It allowed to eliminate tool attachment error which, for such small cutting parameters (a_p, a_e), has a significant impact on the measurement result.

In comparative studies, three carbide milling cutters were mounted in a heat shrinkable holder and three other cutters in 3D printed holders. Specified period of a cutting time and the same cutting parameters were applied for each cutting tool. The first tool ran for 20 min, the second tool for 40 min, and the third tool for 60 min. To simulate multi-axis machining the milling cutters were tilted at an angle of 20 to the spindle axis.

The surfaces obtained during the research were subjected to a qualitative analysis in the form of photos taken with an OLIMPUS microscope, as well as a quantitative analysis with the use of roughness measurement. Figure 1 shows examples of photos of the surface obtained at the beginning of the milling process (first cutting marks) and the surface at the end of the machining process (last cutting marks).

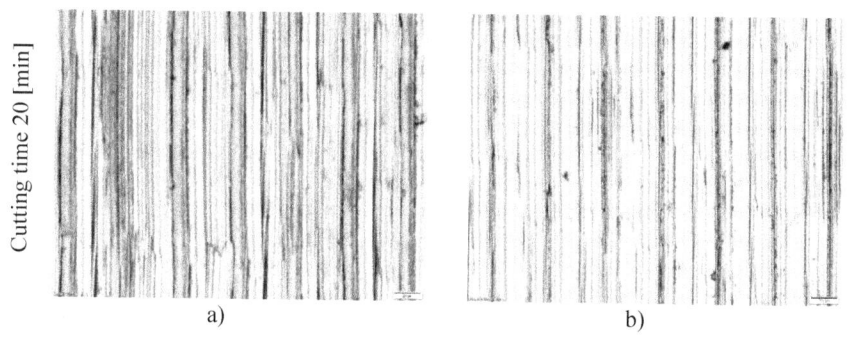

Fig. 1. The views of the machined surface with carbide tools mounted by the 3D printed holder a) initial surface b) the end of cutting process.

It is possible to perform a qualitative assessment of the machined surfaces based on the images (Fig. 1). The machined surface is characterized by clear traces of milling, which is related to the angle of the cutter. The conducted experimental studies of precision milling with carbide tools indicate differences in the quality of the initial and end surfaces of the milled grooves. As the machining time increases, the surface quality improves, which may be related to the lapping process of the cutting-edge during machining.

In the next stage, the impact of the used holder type on the tool wear process was tested. Wear determination is required for the correct and effective production of such components as injection molds. Tool wear measurement was made by 3D scanning of the cutting-edge microgeometry in the selected area and by overlapping the scanned models before and after machining. It should be emphasized that during the tested milling process a very small allowance was removed ($a_p = a_e = 0.04$ mm) by a spiral groove cutter inclined at an angle of 20.

The analysis of 3D scans of carbide cutting edges confirms the stable operation of the tool. During the quantitative analysis of the wear of carbide tools, researchers should only focus on a very small part of the cutting edge. Smaller edge wear for the printed holders can be observed. These values differ very little from the edge wear for heat shrinkable holders. For carbide tools, the maximum cutting-edge wear after 60 min of work for the thermal holder is 0.08 mm and for the printed holder 0.03 mm.

3 Results and Discussion

The properties of the top layer change with the increase of the cutting-edge wear. The assessment of the properties of the surface layer after the long-term operation of the tool may be one of the criteria for the correct or incorrect operation of the tool. In the carried out tests, the influence of the edge wear on the properties of the surface layer was determined. The analysis of the geometric structure of the surface can be carried out on the macroscopic, microscopic and submicroscopic levels. Respectively, waviness, roughness and nanoroughness can be analyzed at these levels. The assessment, of the impact of the cutting-edge wear and the used holder, on the geometric structure of the surface layer was determined on the basis of the conducted experimental tests. The roughness of the obtained surfaces was measured in two directions: in the direction parallel to the feed direction (longitudinal roughness) and in the direction perpendicular to the feed direction (transverse roughness). The research plan and the obtained surface roughness values are presented in Table 2.

Table 2. The surface roughness for the research plan.

Sample	Experimental design					Tool holder	Perpendicular		Parallel	
	a_p [mm]	a_e [mm]	n [rpn]	v_f [mm/min]	Cutting time [min]		Ra [μm]	Rz [μm]	Ra [μm]	Rz [μm]
3	0.04	0.04	9000	1200	20	3D printing	0.192	1.1	0.072	0.389
1	0.04	0.04	9000	1200	40		0.473	2.4	0.157	0.581
5	0.04	0.04	9000	1200	60		0.338	2.1	0.0804	0.4
2	0.04	0.04	9000	1200	20	Heat shrinkable	0.265	1.31	0.0892	0.411
4	0.04	0.04	9000	1200	40		0.428	2.4	0.087	0.368
6	0.04	0.04	9000	1200	60		0.19	1.09	0.0547	0.304

For the surface roughness measurement, a Form Talysurf Intra 50 profilograph was used, with a head and tip for roughness measurement, a scanning table and Talymap Silver and Ultra software. Selected parameters of the tested surface are presented in Fig. 2. The use of the profilometer in the scanning function allowed to obtain isometric views of the tested surfaces. 2D surface roughness was measured and Fig. 3 shows an example of surface roughness profiles.

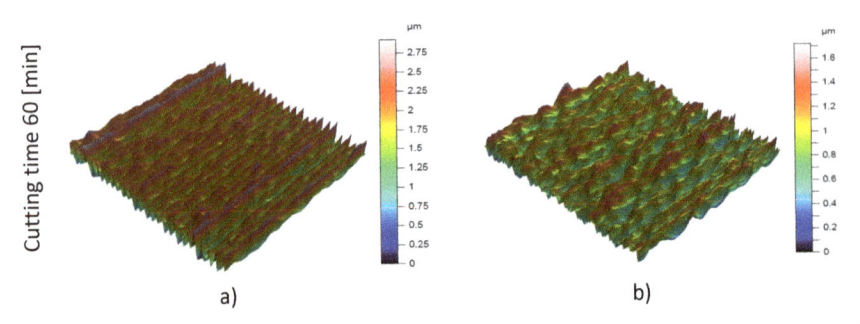

Fig. 2. An exemplary isometric roughness views of machined surfaces for a) the 3D printed holder b) the heat-shrinkable holder.

Impact of the Type of Tool Holder on the Surface Finish   7

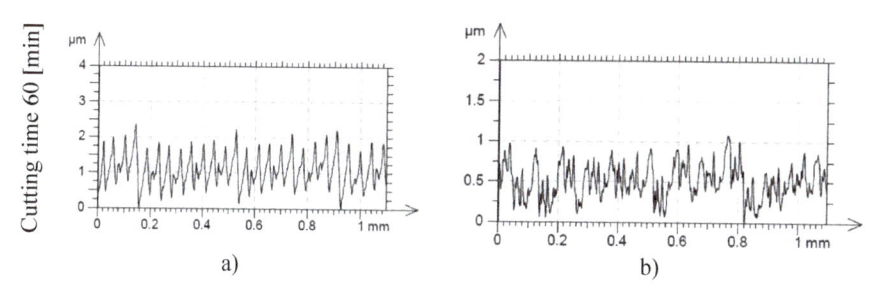

Fig. 3. Surface roughness profiles for a) the 3D printed holder b) the heat-shrinkable holder after 60 min of cutting time.

Operational properties of the workpiece depend on the machining parameters which change the properties of the top layer in a specific way. Long-term operation of the tool may have a negative impact on changes in the properties of the surface layer but also on the metallographic structure of the machining surfaces. In order to check the effect of machining on the structure of the workpiece, the metallographic tests shown in Fig. 4b were carried out. The sample with the lowest roughness values was selected for the tests (sample number 6, tool cutting time 60 min).

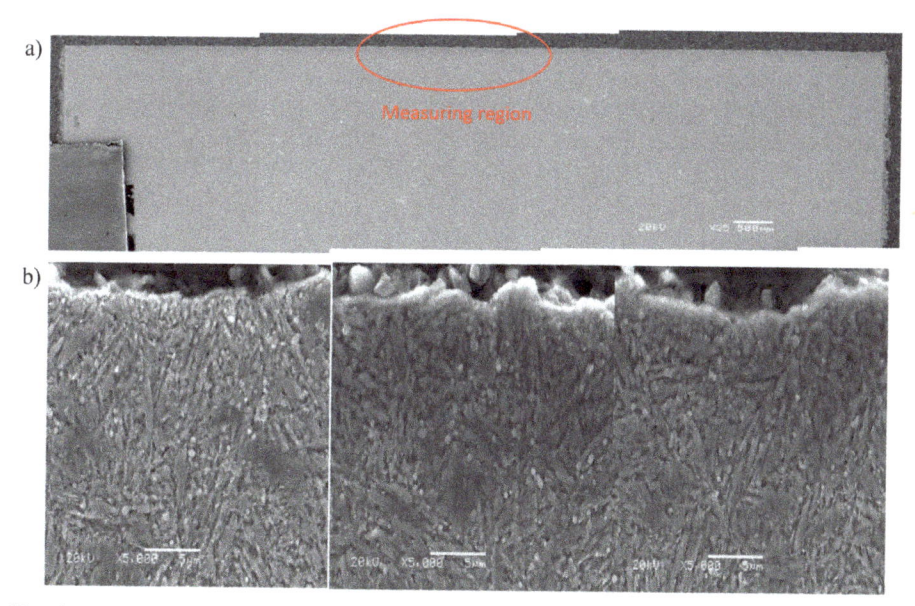

Fig. 4. a) The area of metallographic tests of the sample after milling process b) metallographic examination of the sample from the analyzed area after milling.

Figure 4 shows a photo of the tested sample obtained from the specified measuring region marked in Fig. 4a. The metallographic tests did not show any structural changes in the form of hardening.

4 Conclusions

Wear of the cutting edge starts at the moment of first contact with the workpiece. From the point of view of wear processes, the most dangerous is the moment of cutter entering and exiting from the material. During machining allowance removal process, especially in difficult-to cut materials in the range of 45–65 HRC, the cutting edge works with high mechanical and thermal loads. These working conditions can cause changes in the properties of the tool material, which may cause tool wear and deterioration of its cutting properties. Experimental tests were carried out on the cutters wear with carbide tips during machining of WCLV steel with a hardness of 45–47 HRC. Milling cutters, cutting edge geometry and operating parameters were determined during machine research. The tests were carried out for two different holders: a 3d printed holder and a heat-shrinkable holder. The wear of the carbide tip was very low, which proves stable machining and correctly selected tool operating parameters.

On the basis of the conducted research, it can be concluded that in both cases: for the heat-shrinkable holder and the 3d printed holder, the machining process was carried out under stable conditions.

By analyzing the obtained roughness profile results, for the printed and heat-shrinkable holders, an increase and then a decrease in roughness parameters can be observed with the processing time. The decrease in roughness with the increase in tool working time is related to the lapping process of the tool. During the shaping process, the cutting edge has its own surface roughness, which in the initial stage of the machining process is reduced in the lapping process. The lowest roughness value measured in the perpendicular and parallel direction to the direction of the tool movement was observed for the heat-shrinkable holder and the tool cutting time of 60 min. Similar roughness values were obtained for the printed holder and the cutting time of 20 min.

It can be concluded that for small values of the thickness of the cutting layer, the differences in the surface quality will be similar for both types of holders.

It can be assumed that the advantages of the 3d printed holder, with the structure that affects the vibration damping, will be more noticeable for higher values of the cross-section of the cutting layer.

In order to avoid the formation of vibrations, the conditions for the stable operation of the tool (stable machining) are established. From the point of view of machining, the most important technological parameters that have a significant impact on the stability of the system are the spindle speed and depth of cut a_p. One way to improve machining stability is to reduce the depth of cut. The process parameters tested during machine tests ensured stable machining due to the use of very small cutting depths in the range of 0.02–0.06 mm. The stability of the treatment in the tested range was also confirmed as a result of the surface analysis (Fig. 1).

Acknowledgments. The research was financed as part of the research project POIR.01.01.01-00-0890/17 co-financed by the European Union from the European Regional Development Fund.

References

1. Storch, B., Zawada-Tomkiewicz, A., Żurawski, Ł: Minimal thickness of cut layer in the vicinity of the cutting-edge rounding. Mechanik **10**, 889–891 (2018)
2. Hadia, M.A., Ghania, J.A., Che Harona, C.H., Kasim, M.S.: Investigation on wear behavior and chip formation during up-milling and down-milling operations for Inconel 718. Jurnal Teknologi (Sci. Eng.) **66**(3), 15–21 (2014)
3. Irfaan, M., Temesgen, A., Tsegay, M.: Application of Taguchi method & Anova in turning of AISI 1045 to improve surface roughness by optimize cutting factor. Am. J. Eng. Res. **4**(12), 120–125 (2018)
4. Zhang, J.Z., Chenb, J.C., Kirby, E.D.: Surface roughness optimization in an end-milling operation using the Taguchi design method. J. Mater. Process. Technol. **184**, 233–239 (2007)
5. Irfan, U., Kubilay, A., Fevzi, B.: An experimental investigation of the effect of coating material on tool wear in micro milling of Inconel 718 super alloy. Wear **300**, 8–19 (2013)
6. Krajewska-Śpiewak, J., Gawlik, J.: A method for determination of the minimal thickness of the cutting layer during precision machining performed with the indexable cutting tools. Trans. Tech Publ. Solid State Phenom. **261**, 50–57 (2017)
7. Wojciechowski, S.: Methods of minimum uncut chip thickness estimation during cutting with defined geometry tools. Mechanik **91**(8–9), 664–666 (1018)
8. Wojciechowski, S.: Estimation of minimum uncut chip thickness during precision and micro-machining processes of various materials – a critical review. Materials **15**(1), 59 (2022)
9. Krain, H.R., Sharman, A.R.C., Ridgway, K.: Optimization of tool life and productivity with the use of an end during Inconel 718 milling. J. Mater. Process. Technol. **189**, 153–161 (2007)
10. Murugan Gopalsamy, B., Mondal, B., Ghosh, S., Arntz, K., Klocke, F.: Experimental investigations while hard machining of DIEVAR tool steel (50 HRC). Int. J. Adv. Manuf. Technol. **51**, 853–869 (2010)
11. Tien Dung, H., Nguyen, N.T., Quy, T.D., Thien, N.: Cutting forces and surface roughness in face-milling of SKD61 hard steel. Strojniški vestnik – J. Mech. Eng. **65**(6), 375–385 (2019)
12. Nouari, M., Ginting, A.: Wear characteristics and performance of multi-layer CVD-coated alloyed carbide tool in dry end milling of titanium alloy. Surf. Coat. Technol. **200**, 5663–5676 (2006)
13. Rao, B., Chinmaya, R., Yung, C.: An experimental and numerical study on the face milling of Ti–6Al–4V alloy: tool performance and surface integrity. J. Mater. Process. Technol. **211**, 294–304 (2011)
14. Safari, H., Sharif, S., Izman, S., Jafari, H.: Surface integrity characterization in high-speed dry end milling of Ti-6Al-4V titanium alloy. Int. J. Adv. Manuf. Technol. **78**(1–4), 651–657 (2014)
15. Shokrania, A., Dhokiaa, V., Newmana, S.T.: Imani-Asraia, R., An initial study of the effect of using liquid nitrogen coolant on the surface roughness of Inconel 718 nickel-based alloy in CNC milling. Procedia CIRP **3**, 121–125 (2012)
16. Sonal, C.Y., Rahul, K.S., Mayank, K.M., Pragnesh, K.B.: Investiga-tions on milling tool: - a literature review. Int. J. Res. Eng. Sci. **3**(1), 2320–9356 (2017)
17. Twardowski, P., Hamrol, A., Znojkiewicz, N., Wojciechowski, S.: Analysis of milling cutter working part displacements during milling of steel. Mechanik **10**, 880–882 (2018)
18. Wojciechowski, S., Twardowski, P.: Tool life process dynamics in high speed ball end milling of hardened steel. Procedia **1**, 289–294 (2012)
19. Zhengwen, P., Anshul, S.: High speed ball nose end milling of hardened AISI A2 tool steel with PCBN and coated carbide tools. J. Manuf. Process. **15**(4), 467–473 (2013)
20. Grabowski, M., Gawlik, J., Krajewska-Śpiewak, J., Skoczypiec, S., Tyczyński, P.: Technological possibilities of the carbide tools application for precision machining of WCLV hardened steel. Adv. Sci. Technol. Res. J. **16**(1), 141–148 (2022)

Analysis and Modelling of Intelligent Gate Valve with Fault Detection and Prevention System for Fabrication Using 3D Printing Method

Bushra Mohamad Zakir[1]([✉]), Ahmad Majdi A. Rani[1],
Nu'man Al-Basyir Mohd Nuri Al-Amin[1], Mohd Norhisyam Awang[2],
M. Nasir A. Rani[2], Nur Athirah Ishar[2], Harvey M. Thompson[3], Marek Szostak[4],
and Krishnan Subrmaniam[5]

[1] Department of Mechanical Engineering, Universiti Teknologi PETRONAS, Perak, Malaysia
bushra.mohamad_25297@utp.edu.my
[2] Petronas Chemicals Fertiliser Kedah, Gurun, Kedah, Malaysia
[3] School of Mechanical Engineering, University of Leeds, Woodhouse, Leeds LS2 9JT, UK
[4] Faculty of Mechanical Engineering, Poznan University of Technology, Poznań, Poland
[5] Department of Mechanical Engineering, Manipal International University, Nilai, Negeri Sembilan, Malaysia

Abstract. Gate valves are one of the most extremely used valves in the oil and gas industry. However, previous researchers have found that gate valve tends to fail before its appointed time without prior notice. The causes of failures are often due to many factors which includes internal passing through gate even in fully closed position, wear, and external leakage at packing area. The goal of this study is to design a gate valve with reduced risk of failures and added intelligent functionality which will subsequently lengthen the lifespan of the valve and prevent unnoticed early failures. According to a previous computational study done in 2-dimension (2D), a gate valve's wedge angle is best around 8° at which the fluid inside the valve generates the lowest velocity and pressure and hence the lowest gate deformation. To verify the claim, a 3-dimensional (3D) Computational Fluid Dynamic (CFD) and Finite Element Analysis (FEA) method is conducted to study the fluid flow characteristics and deformation within gate valves in real-world scenario. From the CFD and FEA, the effect of varying wedge gate angle to the velocity, pressure, and gate deformation is studied and areas within gate valve that are most susceptible to failures are identified. Following the analysis, the valve geometry is improved to achieve optimum working conditions and lower risk of failures. In addition to the computational analysis, a study is also made to identify the type of sensor and the practicability of sensor embedded in metal body to add for the intelligent functionality of the valve, so that it is capable of early fault detection. In order to validate the CFD result obtained, a research paper that conducted the 2D analysis is referred to, and the analysis presented is performed identically in 2D via Altair SimLab. The percentage error is between 1 to 5%. Results show that the 3D analysis contradicts previous 2D research that indicated an 8° wedge angle to be the best angle. According to the 3D analysis, the more inclined the wedge angle, the lower the pressure, velocity, and gate deformation, implying that an angle of 9° is better than 8° and an angle of 10° is better than 9°, hence the contradiction. The 3D result is more reliable and accurate as it is more practical in industry.

© The Author(s), under exclusive license to Springer Nature Switzerland AG 2022
M. Diering et al. (Eds.): MANUFACTURING 2022, LNME, pp. 10–25, 2022.
https://doi.org/10.1007/978-3-031-03925-6_2

Keywords: Analysis · Modelling · Intelligent · Gate valve · CFD · FEA · 3D Printing

1 Introduction

1.1 Background

Valves are an essential component of any piping system, particularly in the oil and gas industry. Valves come in varying configurations for various operating needs. They can be categorized specifically into two distinct operations: on/off service and throttling service. On/off valves allow fluid to flow freely with minimal restriction in fully open mode and completely obstructs fluid flow in fully closed position, while throttling valves control flow by restricting or changing the area of the flow path.

Some of the most popular forms of valves used in on/off service are gate valves, plug valves and ball valves. Meanwhile, valves used in throttling include globe valves, diaphragm valves, butterfly valves, and needle valves [1]. Some valves can perform both on/off and regulating functions such as ball valves, butterfly valves, and globe valves [1]. There are also valves designed for special purposes rather than on/off or throttling. Check valves and safety relief valves, for example, are designed to avoid backflow and alleviate pressure in piping system respectively [1].

In this paper, a thorough study will be conducted on gate valve. Gate valves are distinguished by a "gate" which opens and closes to provide full fluid flow through the pipeline or to fully shut off flow. It is used either in the fully closed or fully open positions only. Gate valve should not be used for regulation or throttling of fluid flow since accurate fluid control is not possible. Furthermore, throttling and regulating actions also causes the partially open gate valve to chatter and vibrate, causing degradation (erosion) of the disk and seating surfaces, resulting in internal leakage over time. Throttling of gate valve also creates high noise due to disk chattering thus is impractical to be implemented (Fig. 1).

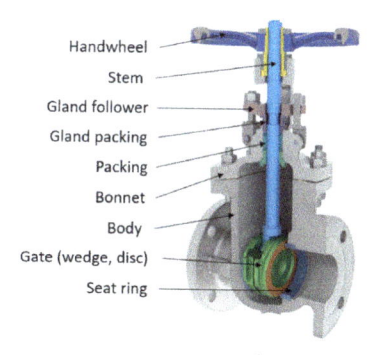

Fig. 1. Gate valve and its internal components.

1.2 Research Problem

The operation of gate valve during on/off application involves the movement of the stem and gate in upward and downward motion. This vertical motion of the gate is controlled by the rotating movement of the handwheel. The rotating motion of the handwheel, translated into the vertical motion of stem, resulting in a slow opening and closing motion. Due to the slow movement of the gate from fully opened to fully closed position, there is high risk of erosion and wear of the gate from the high velocity of fluid flowing through, resulting in the destruction of the tight faces and by time, will lead to passing (internal leakage) due to seal damages. The high pressure generated inside the valve will also cause external leaks through the packing of the valve.

So, an initiative has been made to modify gate valve design to lower the risk of failures and subsequently lengthen the lifespan of the valve. The next generation gate valve will be modified for optimum fluid flow and embedded with sensors capable of alarming early fault detection. Metal 3D printing has been introduced to fabricate the gate valve as it enables the sensors embedding process. The efficiency and the performance are yet to be tested and validation is needed to conduct and prove the success of this next generation gate valve.

1.3 Analysis of Gate Valve

Numerous studies have been performed to improve gate valve functionality and eliminate unnoticed early failures. This can be done by improving the functionality of gate valve which can be further improved if the fluid flow inside them is better understood [2]. To understand fluid flow problems, three approaches can be used which comprises of theoretical approach, experimental approach, and CFD approach [3]. Theoretical approach often requires simplified model of complicated problem which often inaccurate and deviates from actual solution. Experimental approach on the other hand, is not feasible since this project is done via online platform and there is no access to university facilities. Hence, a more reliable and feasible method is used which is CFD method.

CFD approach eliminates the expense of experimental setup and reduces the amount of time taken to achieve solutions via theoretical approach [3]. Moreover, the CFD method is able to return a solution that is similar to the actual result while requiring less time and energy [3]. However, there are some aspects of the CFD approach that should be considered. First, the numerical analysis approach used in CFD is a discrete approximation method, which means that the final solution is derived from a finite number of discrete points and may vary slightly from the actual solution [3]. Second, by the nature of the numerical method, there is no analytical language that can interpret the result in fluid dynamic problems and the conclusions are also nonspecific [3]. Finally, the planning of the software, the selection and use of related resources, and the interpretation of data such as what boundary conditions to be used, are all heavily reliant on the user's knowledge and expertise, which can easily lead to inaccurate results. Nonetheless, the CFD approach remains indispensable across a wide range of engineering issues due to its versatility.

Liu et al. [3] conducted a CFD analysis on subsea gate valve by varying the gate wedge angle and observing the shift in peak velocity and gate deformation. The analysis is done via COMSOL software. From the analysis, he found out that the factor that contributes to the biggest changes in fluid velocity and gate deformation is wedge gate angle 1, α_1, which further followed with gate wedge angle 2, α_2, and wedge gate angle 3, α_3, respectively. The characterization of gate wedge angle is presented in Fig. 2. To demonstrate the robustness of the CFD results conducted by [4], verification is done by reference to experimental results.

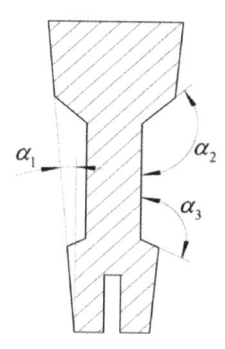

Fig. 2. Description of gate wedge angle 1, 2 and 3 [3].

Zeghloul et al. [4] conducted an experimental investigation on pressure drop inside valve by varying the gate openings. The experiment was conducted on single phase and two-phase flow. The result obtained from single phase flow showed an increase in pressure drop with increasing flowrate and a declination in pressure drops with decreasing gate opening area. Whilst in two-phase flow, in contrast to single phase, the pressure drop was found to increase as the opening area decreases.

In other study performed by Hu et al. [5] on fluid flow pattern inside gate valve using CFD analysis, he found that in the valve upstream, the velocity of the fluid is uniform and stable. However, when it passes the gate with reduced flow area, the velocity increases. Then, the velocity of the fluid abruptly decreases as it flows towards the downstream of the pipeline due to sudden increase in flow area. Hu et al. [5] also stresses on the accuracy of result with different turbulence model. The most accurate analysis result based on his study is the one computed using realizable k-ε and SST k-ω turbulence model. However, SST k-ω took too much time to converge the solution.

1.4 Analysis on Intelligent Functionality

In order to incorporate intelligent functionality into gate valve, many research and previous studies have been studied and reviewed. Based on many researchers, it has been found that one of the most common technology used to detect valve leakage is by using Acoustic Emission (AE) signals. It is a type of non-destructive test (NDT) conducted to measure valve leakage even when the signal is small [6]. Most notably, the valve's operation is not hindered by the operation of this Acoustic Emission (AE) sensor which makes it a superior technique to detect valve leakage.

Lee [7] conducted an experiment to study the capability of using acoustic emission system to detect check valve failures. He uses three types of sensors to accomplish this experiment which is the accelerometer, ultrasonic sensor, and Acoustic Emission (AE) sensor. The accelerometer is used for the detection of mechanical vibration of the valve. The ultrasonic sensor, on the other hand, is used to track disc anomalies, such as when the disc should be completely closed but is not due to a jammed or missing disk, and finally, the Acoustic Emission (AE) sensor is used to localize signals emanating from two distinct failure modes which is the wear of disc, and existence of foreign object. In his study, Lee [7] has succeeded in localizing different type of failures by analyzing and integrating the parameters from the AE signals.

Banjara et al. [8] also carried out an experiment to examine the feasibility of using Acoustic Emission (AE) sensor to detect pipeline leakage. The study is conducted by manipulating the rate of leakage through a pressure-relief valve with seven Acoustic Emission (AE) sensor positioned along the pipeline at a constant distance apart. For different rate of leakage, the seven AE sensor each captures different result due to the difference between the location of sensor and the location of leakage. Results from the study proves that Acoustic Emission (AE) sensor is capable of detecting pipeline leakage by measuring the leakage signals and extracting them for machine learning.

Another experiment conducted by Ye et al. [9] to study the characteristic of valve leakage through Acoustic Emission (AE) signal also prove that Acoustic Emission (AE) sensor is capable of analyzing valve leakage. According to Ye et al. [9], leakage of valve are often due to deformation and wear as a result of corrosion and vibration. This leak would create a void, and the medium in the valve body will eject at a high rate from the leak hole due to the large pressure differential. This vibration coming from the high speed of fluid escaping from the leak can be detected by AE sensor, and analyzed, to obtain detailed result.

2 Material and Methods

The methodology performed to accomplish this project is outlined in this section. There are three primary approaches for this project: Geometry modelling, Computational Fluid Dynamics (CFD), and Finite Element Analysis (FEA).

For this project, three models of wedge gate valve are simulated. The parameter that is observed are the effect of varying gate openings and wedge gate inclination angle to the velocity, pressure, and displacement. The varying gate openings are 10 mm and 1 mm and the varying wedge gate inclination angle are $5°$, $8°$ and $11°$.

2.1 Geometry Modeling

The geometry of the gate valve is adhered to the requirement stated in the API 600 [10]. API 600 [10], with reference to other ASME standards which is the ASME B16.5 [11], ASME B16.34 [12], and ASME B16.10 [13], are used to model the valve. Noted that the gate valve used in this study is wedge gate valve with $2''$ size, 300 Class, Raised Face (RF), and Bolted Bonnet (BB) configurations. Also, the maximum working pressure of the gate valve is obtained from ASME B16.34 [12] by assuming working temperature of 30 °C and material of valve as A216 Gr. WCB.

2.2 CFD Workflow

CFD analysis can be categorized into three main stages namely the pre-processing stage, solver, and post-processing. Pre-processing stage is where the fluid domain within the gate valve is established. Within this stage, the boundary conditions and fluid properties are determined, and geometry mesh are generated. The solver stage then performs numerical calculations to numerically solve the fluid flow equation, and finally, the simulation result obtained from the solver can be visualized using post-processing tools. For this project, Altair SimLab software is used at the pre-processing and post-processing stage while Altair AcuSolve as the solver.

Fluid Domain. The fluid used in this study is methanol at 30 °C with a density of 782 kg/m^3 [14], a dynamic viscosity of 0.000152 Pa.s [14], a specific heat of 2530 J/(kg.K) [14], and conductivity of 0.203 W/(m.K) [14].

Meshing. The meshing size is determined through mesh independence study. An initial mesh size of 1 mm at the edges is simulated, followed by 0.8 mm and 0.4 mm. The reason why the mesh is only refined at edges is because that is the area of interest for this study. This mesh independence study is performed on 5° wedge gate valve with 10 mm opening. The results are tabulated below.

Table 1. Mesh independence study.

Mesh size at edges	Total elements	Computational time	Peak velocity (m/s)	Error
1 mm	689,917	1 h, 2s	21.536 m/s	N/A
0.8 mm	783,969	1 h, 10 min, 50s	21.963 m/s	1.94%
0.4 mm	1,263,927	1 h, 53 min, 2s	22.089 m/s	0.57%

Base on Table 1, mesh size of 0.8 mm is chosen as the error between 0.8 mm and 0.4 mm is only 0.57% which is less than 1%. Moreover, the computing time for 0.8 mm is far lesser than 0.4 mm. Hence, considering the computing time, 0.8 mm mesh size is more convenient and appropriate. The mesh of 0.8 mm at edge is presented in Fig. 3 below.

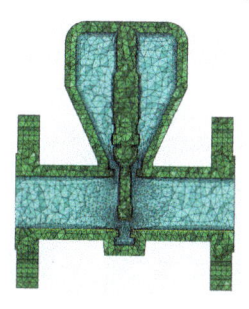

Fig. 3. Meshing of the fluid domain and valve geometry.

Boundary Condition. The boundary condition such as inlet velocity is decided by referring to Methanol Safe Handling Manual [15] which stated that the maximum pipe velocity for the transfer of gasoline/methanol is 23 ft/s which is equivalent to 7 m/s. Hence, a value lesser than 7 m/s should be chosen as the inlet velocity. For this analysis, Standard $k - \varepsilon$ turbulence model is used for solving considering the rapidity and convergence of the solution. The velocity inlet is 5 m/s and the pressure outlet is 0 MPa assuming the outlet is exposed to atmosphere. As for the outer wall of the valve, convection with heat transfer coefficient of 100 W/(m².K), atmospheric temperature of 298 K and valve temperature of 293.15 is used.

Validation of CFD Method. To validate the CFD result obtained, a research paper that conducted a similar analysis is referred to, and the analysis presented is performed identically. Obtaining a similar outcome with previous conducted research would demonstrate that our approach to the problem is on the right track. For this validation stage, CFD analysis conducted by Liu et al. [3] using COMSOL software on subsea gate valve is chosen as reference. Using the same fluid properties, boundary conditions, and valve geometries obtained from Liu et al. [3], the analysis is conducted using Altair SimLab software. However, it is noted that the simulations made by [3] are of 2D geometries. In Altair SimLab, it is not possible to simulate a 2D geometry, hence a constant cross-sectional area is made in Altair SimLab to create similar effect to 2D geometry. An eye to eye comparison is made from the result that is generated and the results obtained from Liu et al. [3]. Based on the analysis on 110 mm stroke, peak velocity was found to be 17.6 m/s while Liu et al. [3] obtained a velocity of 17.4 m/s. The deviation of the result is compared by calculating the % error (Table 2).

$$\%error = \left| \frac{Theoretical - Experimetal}{Theoretical} \right| \times 100\% \quad (1)$$

Table 2. Percentage error between simulated result and results from Liu et al. [3].

Stroke	Simulated peak velocity	Actual peak velocity from [3]	Percentage error
110 mm	9.21 m/s	8.74 m/s	5.4%
120 mm	17.6 m/s	17.4 m/s	1.1%

The calculated value of percentage error is between 1% to 5% which is considered low. Hence, it is proven that our approach to the solution is correct. Figure 4 below shows the velocity profile generated from the analysis in comparison from Liu et al. [3].

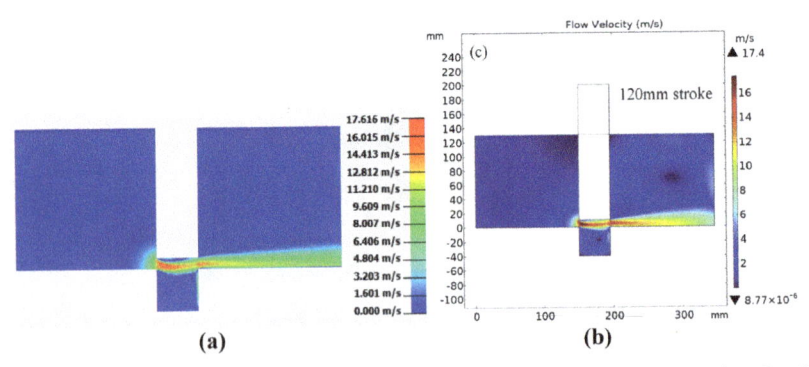

Fig. 4. Velocity profile at 120 mm gate stroke (a) Simulated peak velocity (b) Actual peak velocity from Liu et al. [3].

2.3 FEA Workflow

After finishing the CFD analysis, the temperature and pressure obtained from the CFD analysis are mapped into the structural analysis of the valve. For the analysis, the faces of the bolt holes at flanges of the valve are assigned as fixed support. The bolt holes for fixed support are highlighted as in Fig. 5.

Fig. 5. Faces assigned as fixed support.

The material properties of A216 Gr. WCB for the FEA analysis is obtained from [16] with density of $7800 \, \text{kg/m}^3$, Young's Modulus of $1.9\,\text{E}+11\,\text{Pa}$, and Poisson's ratio of 0.29. For FEA, the same software is used in the pre-processing and post-processing stage meanwhile the solver used is different which is OptiStruct.

3 Results and Discussion

3.1 CFD

The fluid flow pattern is analyzed in wedge gate valve from fully open position (0 mm stroke) to 1 mm to closure (49 mm stroke). The result is plotted in Fig. 6 below.

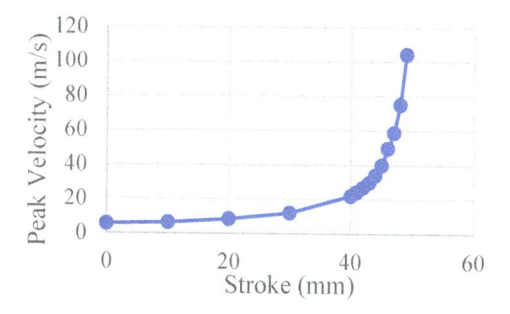

Fig. 6. Graph of peak velocity vs stroke for 5° wedge angle gate valve.

Based on the graph in Fig. 6, it is found that the velocity that passes through the gate decreases with increasing stroke. The graph also shows that the velocity increases dramatically from 40 mm stroke to 49 mm stroke, with the 49 mm stroke having the highest velocity.

Due to the aforementioned reason, it is very crucial in the operation of gate valves that the operator avoids from halting the movement of the gate in midst of closing or opening the gate, particularly during the 40 to 49 mm stroke. This is to avoid the gate from being subjected to extremely high velocity for an extended length of time, which could cause excessive wear thus inflicts failure before its time.

Next, a CFD simulation is made to verify the claim from Liu et al. [3] which stated that an optimal gate valve wedge angle is 8°. Liu et al. [3], in his study, has found that velocity, pressure, and gate deformation is minimum at 8° wedge inclination. However, based on the simulation made through Altair SimLab, the results differ in such a way that the higher the wedge angle, the lower the peak velocity and pressure. The findings from Altair SimLab for gate at 40 mm stroke are presented in Fig. 7. Meanwhile, Fig. 8 and 9 shows the proof of result for velocity and pressure distribution respectively.

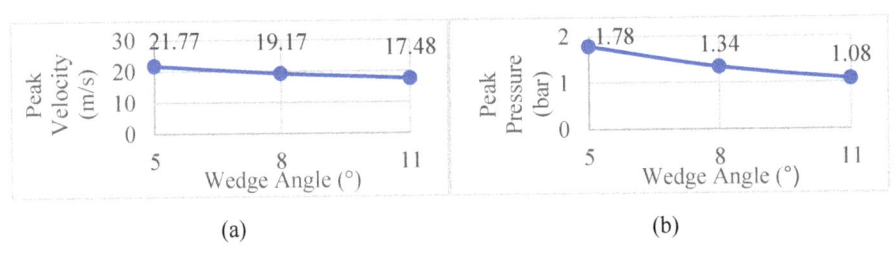

Fig. 7. Graph of (a) Peak velocity (b) Peak pressure vs wedge angle at 40 mm stroke.

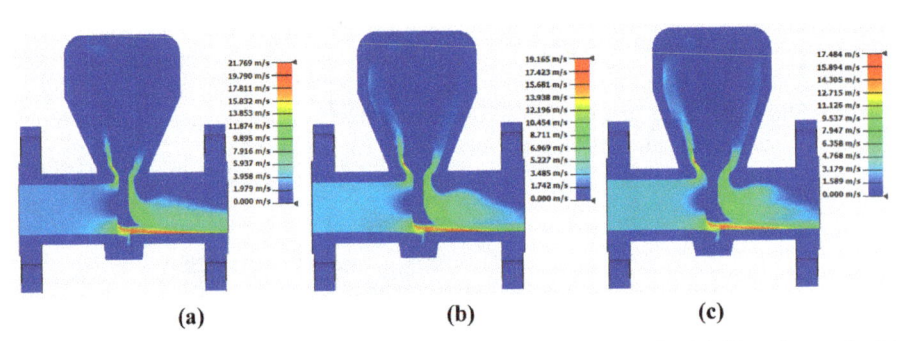

Fig. 8. Velocity distribution at 10 mm valve opening (40 mm stroke) for (a) 5° wedge angle (b) 8° wedge angle and (c) 11° wedge angle.

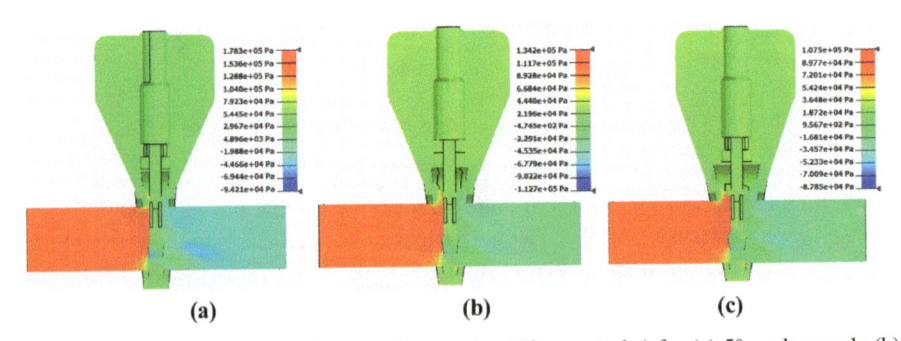

Fig. 9. Pressure distribution at 10 mm valve opening (40 mm stroke) for (a) 5° wedge angle (b) 8° wedge angle and (c) 11° wedge angle.

The simulation is continued for gate valve of 1mm opening (49 mm stroke) since peak velocity occurs here. This time, the difference in velocity and pressure between different wedge angle is quite big. However, the trend remains the same: the more inclined the angle, the lower the velocity and pressure. Again, the claim from [3] failed to be proven. The results are presented in Fig. 10.

In addition, it is found that the peak pressure at 49 mm opening for 5° wedge angle exceeds the acceptable pressure limit for class 300, 2″ valve, standard class, with material A216 Gr. WCB and 30 °C working temperature. The limit for the aforesaid specification is 51.1 bar while the peak pressure generated is 61.9 bar. The result that exceeded the limit is highlighted in yellow in Fig. 10b. This result indicates that a velocity of 5 m/s is not suitable for a 5° wedge angle gate valve as it exceeded the allowable pressure limit at 49 mm stroke.

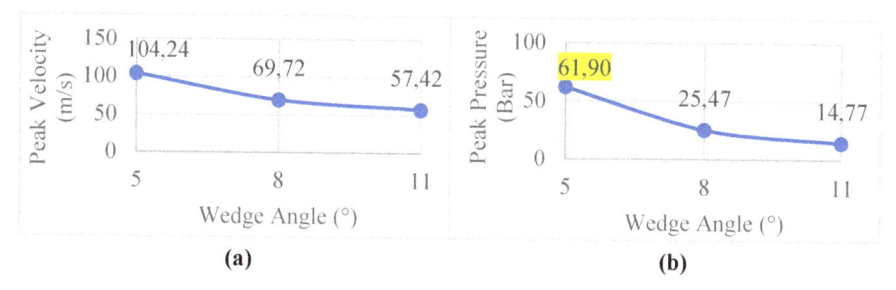

Fig. 10. Graph of (a) Peak velocity (b) of peak pressure vs wedge angle at 49 mm stroke.

In order to substantiate the assertion stated in [3], a further in-depth investigation is conducted to determine the cause of the disparity in results obtained. Figure 11a depicts a gate valve with a welded-in seat ring that is oriented 90° from the gate, while Fig. 11b depicts a valve with a built-in seat ring that is oriented 82° from the gate. Both Fig. 11a and Fig. 11b is of an 8° wedge angle gate valve with different configurations of seat ring.

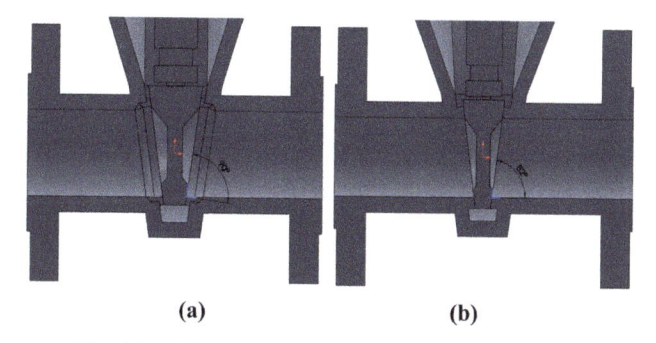

Fig. 11. (a) Welded-in seat ring. (b) Built-in seat ring.

For the record, all our previous simulations used built-in seat rings, which implies that instead of being inserted into the valve body by means of weldment or screw thread, all seat rings are integrated into the valve body. It is suspected that the discrepancies in our result with [3] is due to the difference in orientation of seat ring used in both our study. Hence, a new set of CFD simulation that uses welded-in seat ring is created to observe the changes in fluid flow pattern. To prove that the assumption is correct, a new set of CFD simulation are made to study the fluid flow pattern inside all the three-wedge angle: 5°, 8°, and 11° valve, with welded-in seat rings. The results are as in Fig. 12.

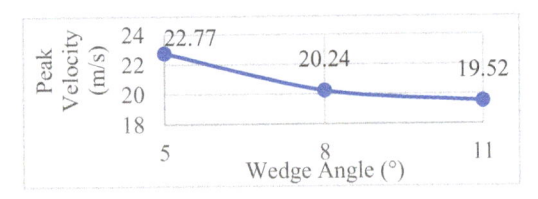

Fig. 12. Peak velocity vs Wedge angle at 40 mm stroke for welded-in seat ring.

The results for the welded-in seat ring showed that the steeper the angle inclination, the lower the peak velocity, which followed the same trend as the built-in seat ring results.

The results obtained still fail to validate the findings discovered by [3] which stated that 8° wedge angle produces minimum peak velocity compared to 5° and 11°. It can be concluded that the simulation made by [3] is inaccurate to be referenced in our study as it utilizes 2D simulation instead of 3D simulation. By using 2D simulation, many of the cross-sectional areas of the valve are disregarded which yields inaccurate result.

3.2 FEA

FEA results are generated for 5°, 8°, and 11° wedge inclination at 49 mm stroke as it is the region that yields the highest peak velocity and pressure. Based on the results in Fig. 13, it was found that the steeper the wedge angle, the lower the deformation of the gate. These results are crucial to study the behavior of the gate when subjected to peak velocity and pressure during the action of opening and closing the valve (Fig. 14).

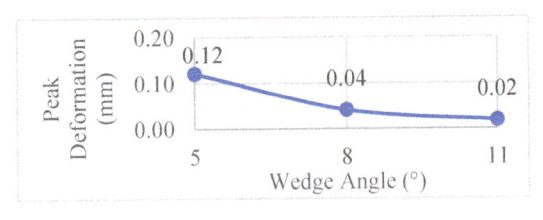

Fig. 13. Graph of peak deformation vs Wedge angle.

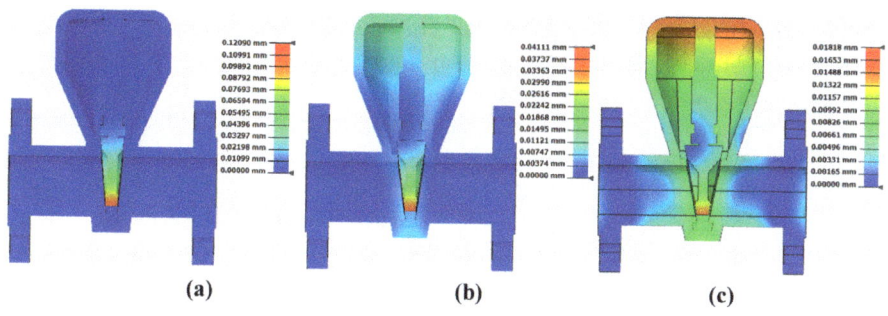

Analysis and Modelling of Intelligent Gate Valve   21

Fig. 14. Gate deformation at 1 mm opening (49 mm stroke) for (a) 5° wedge angle (b) 8° wedge angle and (c) 11° wedge angle.

3.3 Final Geometry

Based on the CFD and FEA analysis generated, it is known that the steeper the angle, the better the performance as it produces lower velocity, pressure, and gate deformation. However, an experimental study should be further made to assess how steep a gate can be made without compromising the practicability of the gate valve in terms of shape, weight, maintenance, etc. Figure 15 below is the geometry of an 8° wedge angle gate valve assuming that it is an optimal angle. Even though the 11° wedge angle gate valve produces the lowest peak pressure and velocity, through naked eyes observation of the 3D model, its gate shape seems bulky which may make it heavier and more difficult to lift. However, as stated earlier, a further investigation should be made to look into this matter.

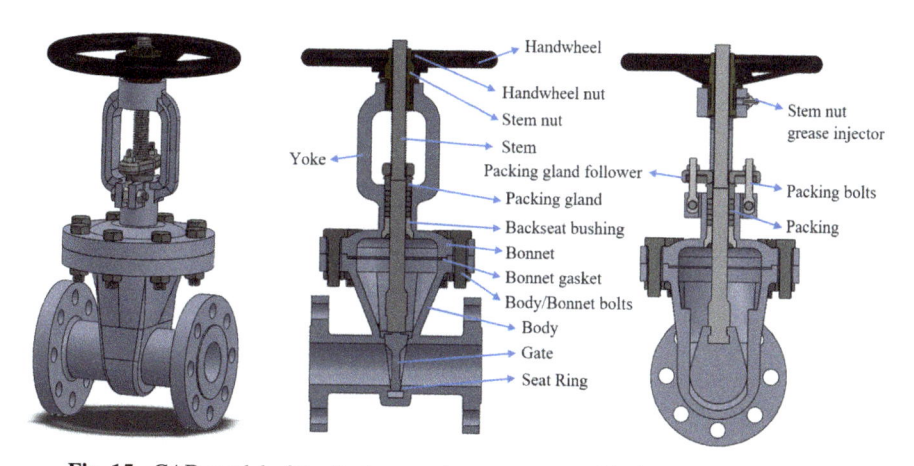

Fig. 15. CAD model of the final gate valve geometry and its internal components.

3.4 Sensor Type and Placement

As discussed in Sect. 1.4, the most suitable sensor that could detect early failures due to leakage and passing is Acoustic Emission (AE) sensor. In industry, valves that leak internally (passing) can result in major losses of valuable product or unintended transfer of process constituents [17]. These can cause process upset due to contamination or even unwanted reaction leading to explosions or fire [17]. In addition, passing could also cause personnel injury if workers assume the upstream pipe is holding (not passing) and open piping for maintenance or repair [17].

By integrating an AE sensor into the valve body, these failures could be detected earlier before it happens. This is because, when leaks or passing occurs, they create gaps or openings to the gate valve that is supposed to be tightly closed, creating a small flow passage that led to the flow of fluid through it. The fluid that passes through the small passage will then create high velocity turbulent flow, which generates solid particles that are impacting on the valve body [8]. Due to the impact, elastic waves are generated and can be detected by AE sensor [8].

Early detection of gate valve passing, or leakage could prevent many disastrous accidents or losses. Today, there is absolutely no way of detecting potential of valve passing before it happens. Thus, by embedding this AE sensor into the valve body, valve passing, and leakages could easily be detected prior to its happening and preventive measures could be properly scheduled.

A study made by [18] has found that wireless AE sensor is indeed possible even though it is not commercialized. The components needed to build this wireless AE sensor are AE sensor module, amplifier, microcontroller, wireless converter, and monitoring system [18]. Firstly, the AE sensor converts the mechanical vibration propagated through the material into electrical signal [18]. The signal will then be amplified by the amplifier, making it usable by the specific software stored in the microcontroller memory [18]. Then, the microcontroller type PIC 18F452, the central unit of the module, commands the amplifier gains and make appropriate analogue-digital conversion [18]. Lastly, the output signal form microcontroller is converted by wireless converter and transmitted in air by antenna (433 MHz) to another antenna (433 MHz) connected to monitoring system at a distance [18]. Figure 16 below shows the wireless AE sensor module.

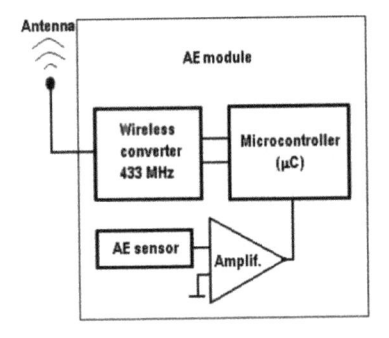

Fig. 16. Wireless AE sensor module [18].

Sensor placement in the valve body needs to be carefully decided as to maintain valve's reliability and integrity. After much consideration, it is decided that sensor placement should be at downstream of the valve as in Fig. 17a. To ensure that the standard thickness of the valve is not compromised, an extra thickness to the valve body is added as in

Fig. 17b. This sensor's location is ideal because it is close to the gate and located at downstream of the valve, where passing frequently happens.

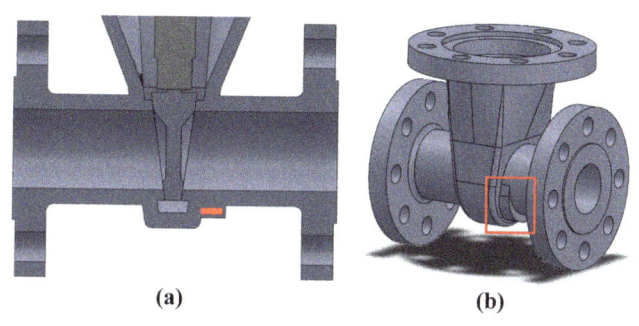

Fig. 17. (a) Sensor position in valve's body. (b) Extra thickness to the valve body to accommodate sensor.

3.5 Method of Intelligent Valve Prototyping

In conventional technology, valves are manufactured by casting method. This makes it impossible to embed sensor into the valve's body. However, today's technology of metal 3D printing which is the Selective Laser Melting (SLM) technology, enables user to produce even the most complex geometry part where there are practically no constraints in fabrication of any complex shaped geometries. SLM works by solidifying metal powder through laser beam, layer by layer according to the geometry of the model. After every solidified layer, a new layer of metal powder is put on top of the preceding hardened layer. The laser beam(s) can be directed and focused through a computer-generated pattern by carefully designed scanner optics [19]. Therefore, the powder particles can be selectively melted in the powder bed and form the shape of 3D objects according to the CAD design of the valve prototype [19].

4 Conclusion

This paper summarizes previous articles and studies on gate valves. CFD and FEA analysis are also carried out. Based on CFD and FEA, the causes of gate valve failures are addressed which is generally due to high velocity of fluid near gate closing state. The velocity, pressure, and gate deformation were found to reach their maximum value at 49 mm stroke of the gate which is 1mm approaching fully closed state. This high velocity fluid will subsequently create higher pressure and increase gate deformation. By modifying the wedge angle of the valve, a significant reduction in velocity and pressure is obtained hence lowering the risks of internal leaks (passing), wear, and external leakage which will subsequently lengthen the lifespan of the valve. It was found that the steeper the angle, the lower the peak velocity, pressure, and gate deformation. Valve's improvement and enhancement was planned according to the causes of failures.

An 8° wedge angle gate was chosen for the improvement as it seems less bulky when observed with naked eyes hence making it easier to lift and operate compared to an 11° wedge angle gate, despite the fact that 11° yields lower peak values. However, further experiments should be conducted to investigate the bulkiness of 11° wedge angle gate compared to 8° to validate the previous statement. Serving its purpose as an intelligent gate valve, a wireless AE sensor will be embodied into the valve's body and fabricated using SLM 3D printing technology.

References

1. Stewart, M.: Piping system components. Surface Production Operations, Elsevier, pp. 193–300 (2016)
2. Mirshamsi, M., Rafeeyan, M.: Dynamic analysis and simulation of long pig in gas pipeline. J. Nat. Gas Sci. Eng. **23**, 294–303 (2015). https://doi.org/10.1016/j.jngse.2015.02.004
3. Liu, P.: Design optimization for subsea gate valve based on combined analyses of fluid characteristics and sensitivity. J. Pet. Sci. Eng. **182**, 106–277 (2019). https://doi.org/10.1016/j.petrol.2019.106277
4. Zeghloul, A., Bouyahiaoui, H., Azzi, A., Hasan, A.H., Al-Sarkhi, A.: Experimental investigation of the vertical upward single- and two-phase flow pressure drops through gate and ball valves. J. Fluids Eng. Trans. ASME **142**(2), 1–14 (2020). https://doi.org/10.1115/1.4044833
5. Hu, B., Zhu, H., Ding, K., Zhang, Y., Yin, B.: Numerical investigation of conjugate heat transfer of an underwater gate valve assembly. Appl. Ocean Res. **56**, 1–11 (2016). https://doi.org/10.1016/j.apor.2015.12.006
6. Li, Z., Zhang, H., Tan, D., Chen, X., Lei, H.: A novel acoustic emission detection module for leakage recognition in a gas pipeline valve. Process Saf. Environ. Prot. **105**, 32–40 (2017). https://doi.org/10.1016/j.psep.2016.10.005
7. Lee, J.H., Lee, M.R., Kim, J.T., Luk, V., Jung, Y.H.: A study of the characteristics of the acoustic emission signals for condition monitoring of check valves in nuclear power plants. Nucl. Eng. Des. **236**(13), 1411–1421 (2006). https://doi.org/10.1016/j.nucengdes.2006.01.007
8. Banjara, N.K., Sasmal, S., Voggu, S.: Machine learning supported acoustic emission technique for leakage detection in pipelines. Int. J. Press. Vessels Pip. **188**, 104243 (2020). https://doi.org/10.1016/j.ijpvp.2020.104243
9. Ye, G.-Y., Ke-Jun, X., Wen-Kai, W.: Standard deviation based acoustic emission signal analysis for detecting valve internal leakage. Sens. Actuators, A **283**, 340–347 (2018). https://doi.org/10.1016/j.sna.2018.09.048
10. Gate, S., Flanged, V., Ends, B., Bonnet, B.: Api 600 (2015)
11. ASME B16.5: Steel Pipe Flanges and Flanged Fittings (2017)
12. ASME B16.34: Valves - Flanged, Threaded, and Welding End (2017)
13. ASME B16.10: Face-to-Face and End-to-End Dimensions of Valves Face-to-Face and End-to-End Dimensions of Valves (2017)
14. Methanol - Thermophysical Properties at Varying Temperatures. https://www.engineeringtoolbox.com/methanol-properties-d_1209.html. Accessed 27 Jul 2021
15. Methanol Institute: Safe Handling of Methnol, p. 149 (2013). http://www.methanol.org/wp-content/uploads/2016/06/Methanol-Safe-Handling-Manual-Final-English.pdf
16. ASTM A216 Grade WCB Cast Steel, MakeItFrom.com. https://www.makeitfrom.com/material-properties/ASTM-A216-Grade-WCB-Cast-Steel. Accessed 27 Jul 2021
17. Passing Valves (leakage) | IPIECA: https://www.ipieca.org/resources/energy-efficiency-solutions/units-and-plants-practices/passing-valves-leakage/. Accessed 27 Jul 2021

18. Chilibon, I., Mogildea, M., Mogilde, G.: Wireless acoustic emission sensor device with microcontroller. Procedia Eng. **47**, 829–832 (2012). https://doi.org/10.1016/J.PROENG.2012.09.275
19. Yap, C.Y.: Review of selective laser melting: materials and applications. Appl. Phys. Rev. **2**(4), 41101 (2015). https://doi.org/10.1063/1.4935926

Effect of Combined Mechanical and Electrochemical Action on Surface Roughness in Microturning Process

Marcin Grabowski[✉], Sebastian Skoczypiec, and Dominik Wyszynski

Chair of Production Engineering, Cracow University of Technology, Kraków, Poland
marcin.grabowski@pk.edu.pl

Abstract. The paper describes the role of electrochemical assistance in microturning process of difficult-to-cut materials. Scaling down a subtractive manufacturing processes introduces many limitations associated with number of physical phenomena and obstacles. The most crucial are forces, allowances and machine tool design as well as cutting tool size and shape. Therefore, it is very important to create appropriate processing conditions enabling optimum results in the form of process flow, parameters, indicators, surface quality and dimensional and shape accuracy of the machined parts and their features. The current research presents basic handicaps appearing in scaling down the microturning process and the results of application of three variants of electrochemical assistance in order to improve size, shape and surface roughness of the machined surfaces. The machining process was investigated for microturning by electrochemical dissolution, microturning by cutting and electrochemically assisted microturning. The presented results confirm important role of hybrid manufacturing due to synergetic effects enabling development and scaling down subtractive processes.

Keywords: Microturning process · Precision machining · Electrochemical assistance · Electrochemical microturning

1 Introduction

In mechanical engineering a concept of miniaturization is related to production of ever smaller parts or their features by using appropriately modified production systems. In industrial practice the basis for classification of manufacturing technologies is a workpiece characteristic dimension. Therefore, methods connected with parts manufacturing where at least one characteristic dimension is less than 1000 μm are used with prefix micro (i.e. microcutting, microturning). The manufacturing in microscale is mainly connected with increase of machining resolution and increase of machine tools and tooling accuracy. Scaling down the turning process means the changes in process and machining characteristics (scale effect occurs) [1]. A decrease of unit removal, that is the reduction of depth-of-cut makes that connection between the cut chip size and the forces of cutting non-linear. The specific energy of cutting also increases (higher energy has to be delivered for material removal). The strength of this result is related to the properties of

© The Author(s), under exclusive license to Springer Nature Switzerland AG 2022
M. Diering et al. (Eds.): MANUFACTURING 2022, LNME, pp. 26–35, 2022.
https://doi.org/10.1007/978-3-031-03925-6_3

machined material (stress and strain), so effective application of turning in microscale is convenient for shaping pieces made of relatively not hard materials only (i.e., mild steel, brass or plastic).

The disadvantageous influence of the size effect could be overpassed if additional energy source was included at machining zone [2, 3]. Developing of a hybrid machining technology by suitable combination of thermal, mechanical and chemical interactions in material removal mechanisms allows obtaining synergetic or economic benefits for the machining. The application of additional energy to the process can improve machinability of the processed material or change the mechanism of material removal. The example of the first strategy for turning in microscale is the thermal-assisted machining [4] while another one can be assisting the cutting process with ultrasonic vibration [5]. However, it is worth emphasising that the solutions mentioned above are not easy to implement in micromachining. Laser assisted machining introduces excessive heat dissipation issue, while vibration of the workpiece or the cutting tool can affect accuracy of the machining. This is why the Electrochemical assistance (ECA) appears to be a noteworthy solution to improve micromachining process.

The idea of combining the electrochemical and mechanical interactions during removal of material was developed in the 1960's as the electrochemical grinding (Abrasive Electrochemical Grinding - AECG) process. Simultaneous mechanical and electrochemical material removal improves the surface layer quality while the tool wear and energy consumption is decreased. The results presented in [6] demonstrate the positive aspect of AECG process assistance for micromachining. Authors concluded thata balance between electrochemical and mechanical means of material removal can be defined with tool feed rate and rotation speed, together with voltage as a set of major parameters for the machining there. Another example is Electrolytic In-Process Dressing (ELID) technology [7], where an electrolytic action (passivation) enables to expose sharp edges of abrasive grains and thus grinding performance is improved. The same solution for improvement of microturning process efficiency was proposed in [8–10]. Concluding from [8], electrochemical passivation of workpiece surface introduces differences in microhardness and topography of the top layer observed after the passivation as compared to prior. In [9, 10] the application of ECM in microturning was investigated. In such a process, depending on ECA parameters material can be removed: (first) with simultaneous mechanical and electrochemical treatment or (second) by application of electrochemical reactions in order to trigger a change of the mechanical conditions for material machining. In both of the above-mentioned cases, electrochemical assistance reduces the mean cutting force value by over a dozen to a few millinewtons, which translates to a reduction of a dozen to several percents (Fig. 1). Additionally–for selected machining parameters–the application of ECA has impact on lower wear of the tool tip and main cutting edge.

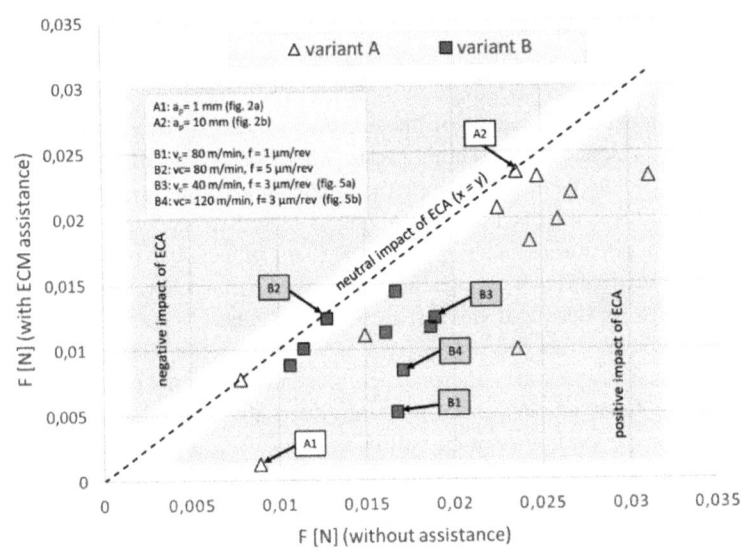

Fig. 1. Correlation between force sensor load F without and with ECA for two following variants of microcutting: variant A - material is removed with simultaneous electrochemical and mechanical action and variant B – ECA is applied to change the mechanical conditions of the machined material; ap – depth of cutting, vc – cutting speed, f – machining feed [10].

2 Research Problem

The electrochemical processing is considered as a process, which allows to obtain workpiece with high quality of the surface layer. Therefore, one can expect, that application of ECA in microturning would allow to improve surface integrity in comparison to the classic (i.e. mechanical) process. It is worth mentioning that in classic process cutting depth is in range of minimal chip thickness and the significant role during the machining plays the ploughing effect. It has a negative effect on surface integrity of microparts machined in such an operation.

The experimental part of this paper consists of research methodology and a comparison of straight turning results in the case of 1.4301 stainless steel machining in following three variants: (A) - microturning by electrochemical dissolute (B) - microturning by cutting and (C) - electrochemically assisted microturning. The analysis of obtained results was focused on workpiece surface roughness after machining.

The concept of ECA for microturning process was verified in straight turning kinematics, where both the cutting process and ECA could act simultaneously. The machining process was investigated for presented in Fig. 2 variants i.e.: microturning by electrochemical dissolution (variant A) microturning by cutting (variant B), and electrochemically assisted microturning (variant C). Variants A and B represent two extreme cases where machining force is equal to zero (A) and maximum value (B). The following conclusions can be drawn:

- microturning by electrochemical dissolution (variant A) is connected with material removal by electrochemical dissolution without additional mechanical or thermal forces and no tool wear occurs. Efficiency and surface quality is much higher than during cutting. The process allows to obtain smaller shaft diameters. However, it's accuracy is limited due to problems with localization of electrochemical reactions,
- microturning by cutting (variant B) is connected with typical scaling problems i.e. increased tool wear, increased specific cutting energy with decrease of cutting depth or workpiece deformation. However, in comparison to variant A micro turning by cutting allows to obtain better accuracy.

The electrochemically assisted microturning (variant C) is connected with simultaneous mechanical and electrochemical removal of the machined material. Solutions like this combine crucial advantages of both processes: high accuracy of mechanical machining with no tool wear and absence of mechanical forces in electrochemical microturning. This solution also simultaneously reduces the major disadvantages of both processes: relatively high specific cutting energy of mechanical microturning and relatively low accuracy of electrochemical microturning.

The test stand for the research presented below was detailed in [9]. The tests described below were planned in accordance with Design of Experiment theory. The input parameters were presented in Table. 1. It is noteworthy that the workpiece diameter was 3 mm, which is out of range of micromachining, however selected depth of cut (<10 μm) is typical for microturning operations. While the main goal of the research was to develop main technological characteristics of the investigated processes [11], only issues of surface roughness will be addressed in the next paragraphs. Surface profile was measured with application of Taylor Hobson Form Talysurf Series 2 device.

Table 1. Input machining parameters during experiments (scheme of investigated variants in Fig. 1).

	Variant A	Variant B	Variant C
Cutting speed V_c [m/min]	30 and 90	3 and 11	40–120
Feed rate, f [μm/min]	(1 and 5 μm/rev)	200–500	(1–5 μm/rev)
Depth of cut, a_p [μm]	1–10		1–10
Initial interelectrode gap thickness, S [mm]		0.005	0.25
Interelectrode voltage, U [V]		10–20 (pulse voltage, $t_i = t_p = 10$ μs)	10 (pulse voltage, $t_i = t_p = 10$ μs)
Workpiece rotation speed, n 1/min	3183–9550	300–1200	4245–12733
Number of machined layers	10	1	3 + 3 (first 3 layers without ECA)

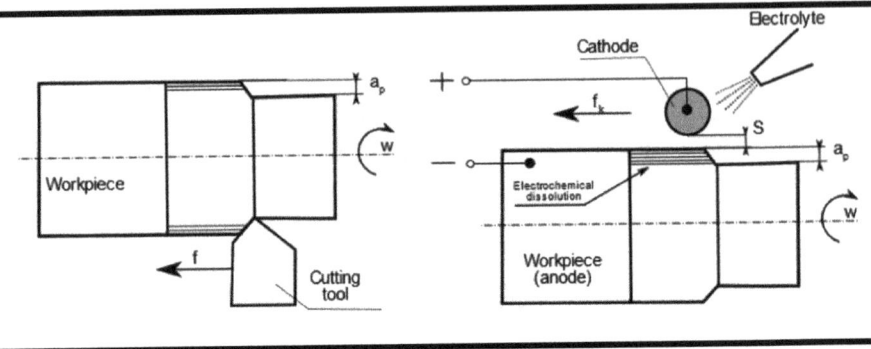

Microturning by cutting (variant A) | **Microturning by electrochemical dissolution (variant B)**

Electrochemically assisted microturning (variant C)

Description of symbols and constant machining parameters:
- w – workpiece rotation
- f – cutting feed
- fk – cathode feed
- S – initial interelectrode gap size
- Workpiece material: 1.4301 Stainless Steel
- Workpiece diameter: 3 mm
- Cutting tool: Mitsubishi Turning Insert DCET070200R-SN NX2525
- Cathode: material – tungsten carbide,
- Cathode Diameter - 0,4 mm
- Electrolyte: 1% $NaNO_3$ deionized water solution
- Length of turning: 4 mm

Fig. 2. Schemes of investigated variants of microturning process.

3 Results and Discussion

3.1 Microturning by Cutting (Variant a)

During the experiment the relevant influence of technological parameters (i.e. cutting speed and tool feed) on surface roughness was found [11]. Application of small value of depth of cuts favors the plastic deformation and ploughing in material removal mechanism, what is especially noticeable for small values of cutting speed. Below only results for ap = 1 μm and extreme values of other investigated parameters will be discussed (Figs. 3, 4 and 5).

a) f = 1 µm/obr. v = 30 m/min

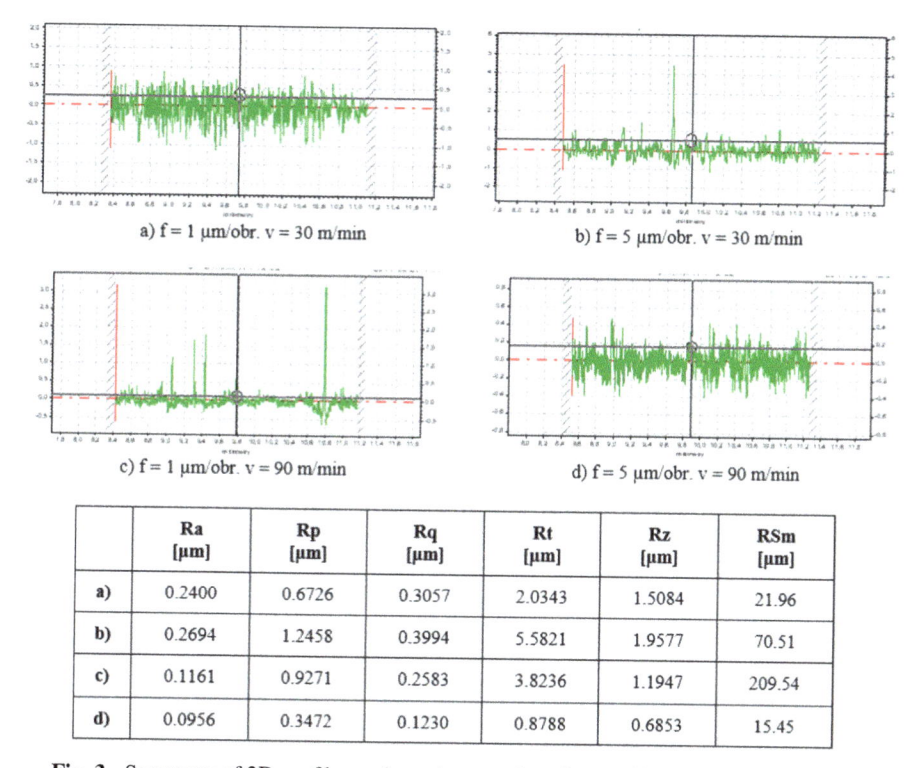

b) f = 5 µm/obr. v = 30 m/min

c) f = 1 µm/obr. v = 90 m/min

d) f = 5 µm/obr. v = 90 m/min

	Ra [µm]	Rp [µm]	Rq [µm]	Rt [µm]	Rz [µm]	RSm [µm]
a)	0.2400	0.6726	0.3057	2.0343	1.5084	21.96
b)	0.2694	1.2458	0.3994	5.5821	1.9577	70.51
c)	0.1161	0.9271	0.2583	3.8236	1.1947	209.54
d)	0.0956	0.3472	0.1230	0.8788	0.6853	15.45

Fig. 3. Summary of 2D profiles and roughness values for machined shafts (variant A).

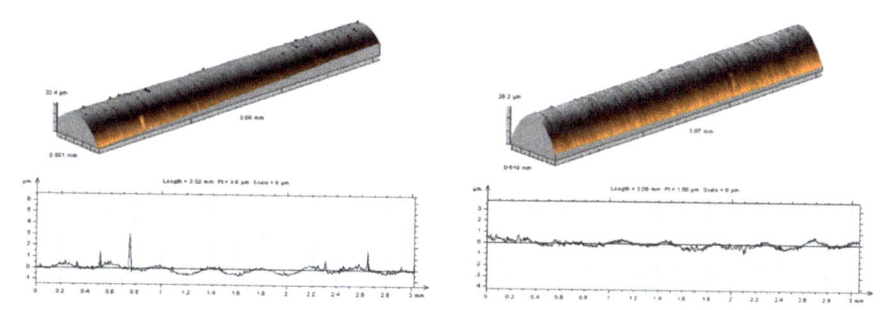

Fig. 4. 3D and 2D profiles of the shaft surface machined in variant A; $f = 1\,\mu\text{m/rev}$, cutting speed $v_c = 90\,\text{m/min}$.

Fig. 5. 3D and 2D profiles of the shaft surface machined with in variant A; feed $f = 5\,\mu\text{m/rev}$, cutting speed $v_c = 90\,\text{m/min}$.

One can state that the surface roughness decreases with increase of cutting speed, while with feed rate increase roughness increases too. The higher cutting speed means more stable process with reduced tool vibrations, however relation of surface roughness with cutting speed relates to mechanical properties of machined material. For small values of cutting speed the plastic properties play important role in material removal mechanism and are reason of built-up edge phenomena. The increase of cutting speed decreases share of plastic phenomena and surface roughness increases. It is worth to mention, that surfaces machined with small values of feed has traces of strong plastic deformation, what indicates occurrence of ploughing phenomena. Probably, for selected machining parameters and feed f = 1 μm/rev, machining take place in range of minimal depth of cut.

3.2 Microturning by Electrochemical Dissolution (Variant B)

The research of electrochemical microturning was carried out according to research plan with five repetitions for central value. The mean and variance value were calculated for whole investigated cases and only for center of the plan were similar and had the following values:

- mean value of Ra = 0.082 μm with variance 0.012 μm,
- mean value Rz = 0.545 μm with variance 0.1 μm.

Total correlation coefficient R between U, F and n and values of Ra and Rz was 0.36 what indicates poor relation of surface roughness to process parameters. It is worth to mention, that selected parameters of electrochemical microturning was typical for finishing. The electrochemical process carried with small interelectrode gap is connected with high current density (in investigated case it was in range 400–700 A/cm^2) and results in high quality of surface layer (examples presented in Figs. 6 and 7) for full range of investigated parameters. The obtained in electrochemical microturning surface roughness varies slightly and is not related to the process efficiency.

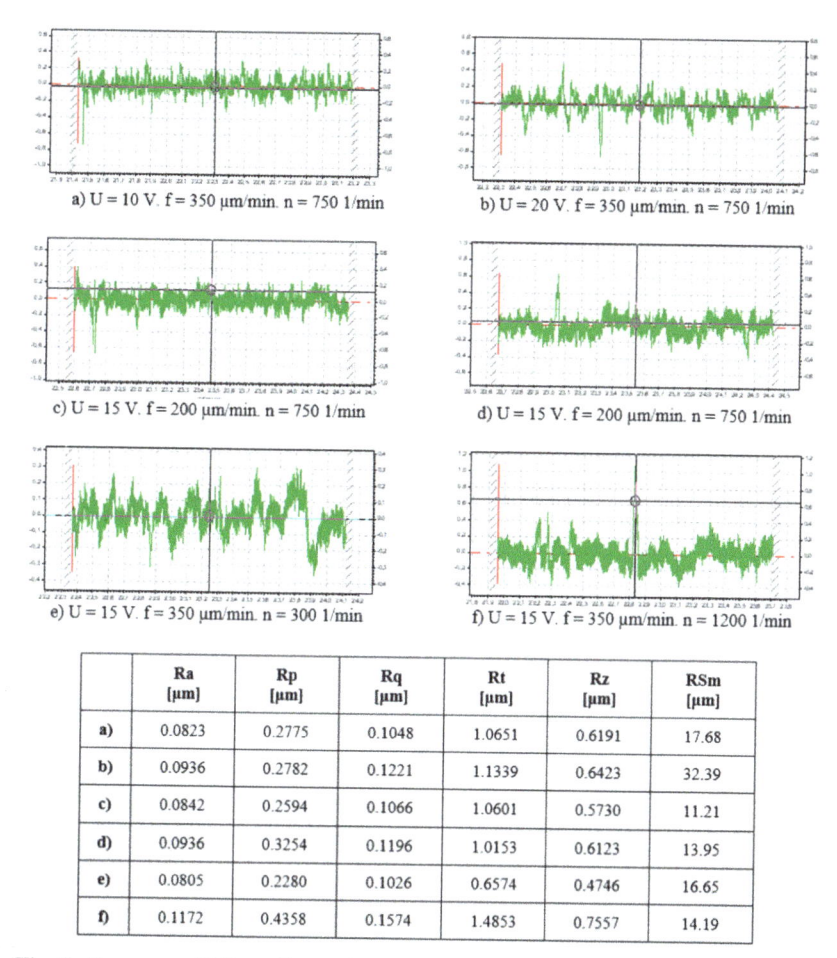

Effect of Combined Mechanical and Electrochemical Action 33

a) U = 10 V. f = 350 µm/min. n = 750 1/min

b) U = 20 V. f = 350 µm/min. n = 750 1/min

c) U = 15 V. f = 200 µm/min. n = 750 1/min

d) U = 15 V. f = 200 µm/min. n = 750 1/min

e) U = 15 V. f = 350 µm/min. n = 300 1/min

f) U = 15 V. f = 350 µm/min. n = 1200 1/min

	Ra [µm]	Rp [µm]	Rq [µm]	Rt [µm]	Rz [µm]	RSm [µm]
a)	0.0823	0.2775	0.1048	1.0651	0.6191	17.68
b)	0.0936	0.2782	0.1221	1.1339	0.6423	32.39
c)	0.0842	0.2594	0.1066	1.0601	0.5730	11.21
d)	0.0936	0.3254	0.1196	1.0153	0.6123	13.95
e)	0.0805	0.2280	0.1026	0.6574	0.4746	16.65
f)	0.1172	0.4358	0.1574	1.4853	0.7557	14.19

Fig. 6. Summary of 2D profiles and roughness values for machined shafts (variant B).

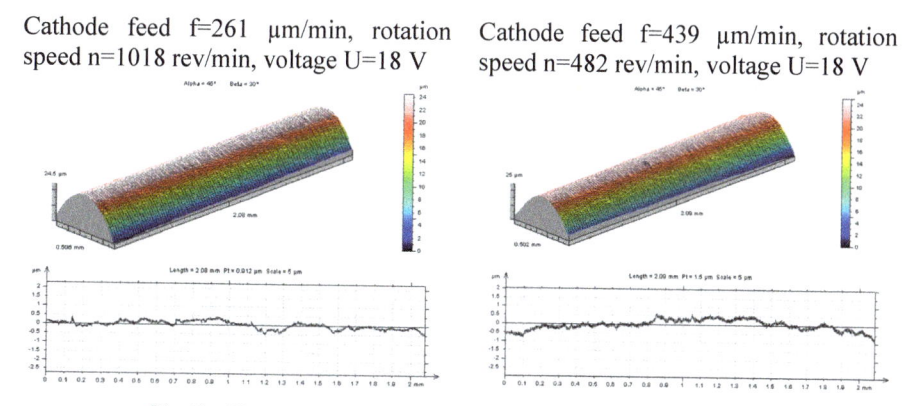

Cathode feed f=261 µm/min, rotation speed n=1018 rev/min, voltage U=18 V

Cathode feed f=439 µm/min, rotation speed n=482 rev/min, voltage U=18 V

Fig. 7. 3D and 2D profiles of shaft surface machined in variant B.

3.3 Electrochemically Assisted Microturning (Variant C)

According to the results presented in [10, 11] for investigated range of cutting parameters the electrochemical assistance gave the most benefits for ap = 1 µm. Increase of ap caused that influence of ECA becomes negligible, while for small values of ap electrochemical dissolution carried simultaneously with cutting decreased width of cutting layer and decreased cutting force. The same effect was noticeable for small values of cutting speed (Vc = 40 m/min), when plastic deformation played important role. In such case ECA improved condition of material removal by change of dislocation density in surface layer.

The measured with Taylor Hobson Surtronic 25 surface roughness for all samples was in range: Ra = 0.08–0.15 µm and Rz = 0.5–1.15 with standard deviation σRa = 0.02 µm a σRz = 0.22 µm. Considering the values of Ra and Rz obtained for classical process (variant A, paragraph 3.1) one can state that in ECA microturning phenomena connected with electrochemical dissolution has dominant influence on the formation of technological surface layer. It is worth mentioning, that for all **variant C** tests the same parameters of ECA were applied and current density was $\approx$10 A/cm^2. Such value was taken as boundary for the correctly designed electrochemical machining process. Figure 8 and 9 shows the surface resulting from the cutting process in variant A and C. An improvement in the machined surface was obtained because of the electrochemical support.

Fig. 8. Comparison of surfaces after micro turning a) without electromechanical assistance b) with electromechanical assistance.

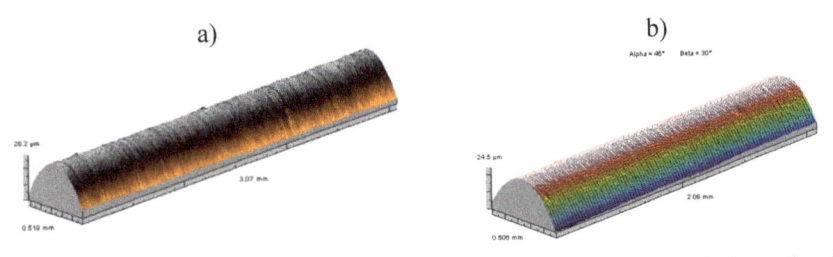

Fig. 9. 3D profiles of the shaft machined in mechanical microturning (a) and electrochemical assisted microturning process (b).

4 Conclusions

Considering the above-mentioned results one can conclude that values of surface roughness after microturning (variant A) are several times higher than in case of electrochemical microturning (variant B) and electrochemically assisted microturning process (variant C). Machining with variant A means, that surface roughness is in strong relation with other machining parameters, while in variant B and C varies slightly. In case of electrochemical - based machining variants surface roughness is not related to mechanical properties of material (scale effect), therefore ensuring of typical for finishing process condition guarantees good surface quality.

The research presented in the paper shows the potential benefits of ECA during cutting. However, it is worth mentioning, that ECA is effective only when the depth-of-cut is ≤ 1 μm, therefore such a solution is not very effective in this case. Accordingly, its application should be considered only at the micropart final stage machining, especially when cutting depth is in the minimal chip thickness range and the ploughing effect plays a main role in material removal mechanism. It is also worth to underline that the proposed solution can be applied to machine only conductive materials.

The presented combined process is promising machining technology, however its further development is connected with explanation of physical nature of synergy between mechanical and electrochemical action. Taking into account mentioned above advantages one can also suggest to carry out research for other variants of machining i.e. grinding.

References

1. Vollertsen, F., Biermann, D., Hansen, H., Jawahir, I., Kuzman, K.: Size effects in manufacturing of metallic components. CIRP Ann. Manuf. Technol. **58**, 566–587 (2009)
2. Lauwers, B., Klocke, F., Klink, A., Tekkaya, A.E., Neugebauer, R., Mcintosh, D.: Hybrid processes in manufacturing. CIRP Ann. Manuf. Technol. **63**, 561–583 (2014)
3. Chavoshi, S.Z., Luo, X.: Hybrid micro-machining processes: a review. Precis. Eng. **41**, 1–23 (2015)
4. Sun, S., Brandt, M., Dargusch, M.S.: Thermally enhanced machining of hard-to-machine materials a review. Int. J. Mach. Tool Manu **50**, 663–680 (2010)
5. Brehl, D., Dow, T.: Review of vibration-assisted machining. Precis. Eng. **32**, 153–172 (2008)
6. Zhu, D., Zeng, Y., Xu, Z., Zhang, X.: Precision machining of small holes by the hybrid process of electrochemical removal and grinding. CIRP Ann. Manuf. Technol. **60**, 247–250 (2011)
7. Rahman, M., Kumar, A.S., Biswas, I.: A review of electrolytic in-process dressing (ELID) grinding. Key Eng. Mater. **404**, 45–59 (2009)
8. Skoczypiec, S., Grabowski, M., Spychalski, M.: Experimental research on electrochemically assisted microturning process. Key Eng. Mater. **611–612**, 701–707 (2014)
9. Skoczypiec, S., Grabowski, M., Ruszaj, A.: The impact of electrochemical assistance on the microturning process. Int. J. Adv. Manuf. Technol. **86**(5–8), 1873–1880 (2016)
10. Grabowski, M., Skoczypiec, S., Wyszynski, D.: A Study on microturning with electrochemical assistance of the cutting process. Micromachines **9**, 357 (2018)
11. Grabowski, M.: Wspomagany elektrochemicznie proces toczenia mikroelementów. Praca doktorska, Politechnika Krakowska, Kraków (2014)

Evaluation of the Fidelity of Additively Manufactured 3D Models of a Fossil Skull

Miroslaw Rucki[1(✉)], Yaroslav Garashchenko[2], Ilja Kogan[3,4], and Tomasz Ryba[5]

[1] Faculty of Mechanical Engineering, Kazimierz Pulaski University of Technology and Humanities in Radom, ul. Stasieckiego 54, 26-600 Radom, Poland
m.rucki@uthrad.pl

[2] Department of Integrated Technologic Process and Manufacturing, National Technical University «Kharkiv Polytechnic Institute», Kyrpychova Street 2, Kharkiv 61002, Ukraine

[3] TU Bergakademie Freiberg, Geological Institute, Bernhard-von-Cotta-Street 2, 09599 Freiberg, Germany

[4] Institute of Geology and Petroleum Technologies, Kazan Federal University, Kremlyovskaya 4/5, 420008 Kazan, Russia

[5] Sieć Badawcza Łukasiewicz, Instytut Technologii Ekspoatacji w Radomiu, ul. K. Pułaskiego 6/10, 26-600 Radom, Poland

Abstract. The paper presents a comparative analysis of the 3D-printed models of a complex geometrical object obtained using different Additive Manufacturing technologies. The object of interest is a unique fossil skull of a 'reptiliomorph amphibian' *Madygenerpeton pustulatum*. Twelve different copies were 3D-printed using the same (reference) digitized model and then scanned with a Mitutoyo Coordinate Measuring Machine (CMM) CRYSTA-Apex S 9166. Fidelity of each copy was assessed through the comparison with the reference digital model and with each other in couples. Statistical analysis of the distances between compared surfaces provided good background for the choice of the most accurate copies.

Keywords: Additive manufacturing · Madygenerpeton · Optical measurement

1 Introduction

Additive manufacturing (AM) technologies are relatively new methods using an incremental layer-by-layer materialization of a digital model. The AM methods referred to also as 3D printing (3DP), rapid prototyping (RP), or solid-freeform (SFF) became an exponentially evolving manufacturing technology [1]. The estimated average global value of the 3D printing market is recording a 25% year-to-year increase since 2014, and is expected to reach 35.0 billion USD by 2024 [2]. A good review of these methods can be found in [3], as well as in more recent works related to Industry 4.0 and Internet of Things concepts [4, 5]. It is emphasized that the cost of Additive Manufacturing is a crucial factor [6]. There are also works that discuss design principles, constrains and optimization for AM techniques [7].

At present, a large variety of individual additive processes are available depending on a material and machine technology, which can be classified into the seven main categories: material jetting, binder jetting, vat photopolymerization, material extrusion, powder bed fusion, sheet lamination, and direct energy deposition [8]. AM allows for hybrid- and multi-material (MM) manufacturing of metals and is especially suitable for functionally graded materials (FGMs) [9]. There are well-established and rapidly emerging applications of AM technologies, such as rapid fabrication of a prototype, micro-scale manufacturing for aerospace and motor industries, medical applications, rapid tooling, direct digital manufacturing, as well as an increasing number of new applications far beyond the initial intent of prototyping [10].

From the literature review it can be concluded that among the main directions of 3D-printing development are geometry and material design for AM, computational tools and interfaces development, as well as manufacturing tools and processes development [11]. In most of published research papers, attention is paid to the optimization tasks of AM-process planning [12], CAD-AM (RP) programs [13], choice of AM technologies [14], the synthesis principles applied during the manufacturing process [15], design of cladding layers, the proper choice of building materials, as well as finally obtained surface quality [16], dimensional accuracy [17], as well as microstructures and properties [18]. The objective of this paper is to assess fidelity of the 3D-printed models of a natural complex geometrical object with no initial documentation, fabricated using different Additive Manufacturing technologies. The object of interest is a fossil skull of a 'reptiliomorph amphibian' *Madygenerpeton pustulatum* [19] from the Triassic of Kyrgyzstan. The skull is somewhat deformed and lacking the lower jaw, but because of its uniqueness it is of great interest to the paleontological community, and its 3D-printed models are intended for exhibitions, teaching and research purposes.

2 Materials and Methods

The replicated object is the holotype of *Madygenerpeton pustulatum* [19], an incomplete fossil skull that was recovered in 2007 from the fossil deposit Madygen in southwest Kyrgyzstan [20], world's arguably richest non-marine finding locality of Triassic fossils (ca. 237 million years before present). This unique skull has been deformed by fossilization processes and lacks the lower jaw. Figure 1 presents the overall view of the fossil and the close-up image of its surface morphology with bony tubercles covering the skull. No complete remains of *Madygenerpeton pustulatum* individual has been found. Bony carapace shields (osteoderms) discovered next to the skull belonged to at least three individuals of the same species. *Madygenerpeton* is a member of the extinct *Chroniosuchia*, a group of derived amphibians close to the origin of all higher vertebrates (mammals and reptiles including birds), and is reconstructed to have been a crocodile-like predatory animal adapted to both terrestrial and aquatic locomotion. It can be assumed that the osteoderms would act as an additional trunk support while walking on land, but at the same time, their weight would push the body below the water surface when dwelling in water [21].

Fig. 1. Fossil skull of *Madygenerpeton pustulatum* (holotype FG 596/V/4, housed at the TU Bergakademie Freiberg): a) view, and b) close-up picture of surface morphology.

To reach the main objective of the research, following steps were undertaken:

- Preparation of the 3D-printed copies of the object using the same digital model (later used as a reference), but different AM techniques,
- Scanning of the physical copies of the object to obtain their digital models,
- Visual and statistical analysis of the differences between the initial (reference) digital model,
- Detailed statistical analysis of the differences in order to determine peculiarities of the applied AM methods and to point out the most suitable one for the investigated *Madygenerpeton* fossil skull.

After scanning of the object with different scanning techniques, rating of the digital models was made [22]. In the rating, the digitization with AICON SmartScan exhibited the lowest distances to the models obtained from other scanners. Nevertheless, for the 3D-printing of the copies, Artec Space Spider digitized model was chosen. It was in the third place of the rating, which can be considered as proof for high accuracy. Moreover, it was advantageous for practical reasons, namely, it provided better the detailed geometry of the smallest elements of the fossil skull. As a result, the digitized 'Artec surface model' was 20,978.512 mm^2, which was by 11.8% larger than that obtained from AICON device. Thus, the reference digital surface in this research was the model obtained from scanning the fossil skull with the Artec Space Spider device.

Next, 12 physical copies of the object were 3D-printed using a wide range of the available AM techniques. Table 1 shows the main data of the devices and parameters used in experiments, while the Fig. 2 presents the photo of model #7 made out of the apricot kernel flour.

The 3D-printed copies were then scanned with a Mitutoyo Coordinate Measuring Machine (CMM) CRYSTA-Apex S 9166 at Mitutoyo Polska, Wrocław. The maximum permissible error of the CMM was $MPE_E = \pm (1.7 + 3L/1000)$ µm. The surface scanning was performed with non-contact line laser probe SurfaceMeasure 606 with scanning error 12 µm [1σ/ sphere fit]. Its accuracy was found satisfactory after preliminary analysis [23]. The scanning procedure was performed in a single fixation in order to minimize the error generated by the formation of a points cloud.

Table 1. Technologies and devices used for manufacture of the fossil skull copies.

Copy No	Technologies	Equipment model	Material	Layer thickness, mm
#1	Extrusion-based FDM	Ultimaker 2 +	PLA	0.100
#2	Multijet fusion	HP jet fusion 3D 4210	Polyamid PA 12-HP	0.080
#3	Extrusion-based FDM	Prusa i3 MK2	PLA	0.050
#4	Powder-based 3D printing with inkjet	Canon ProJet 460Plus	White gypsum	0.125
#5	UV-resin-Inkjet	Continuous inkjet printers keyence	Yelllowish transparent	0.100
#6	Powder-based 3D printing with inkjet	ZCorp 310(R)	Gypsum, ZP151	0.088
#7	Powder-based 3D printing with inkjet	ZCorp 310(R)	Apricot kernel flour	0.088
#8	Polyjet	Stratasys J55	Standard material	0.019
#9	Polyjet	Stratasys J55	Vivid material	0.019
#10	ColorJet printing	3D Systems ProJet CJP 460Plus	VisiJet PXL	0.100
#11	ColorJet printing	3D systems Zprinter 650	VisiJet PXL	0.089
#12	UV-curable inkjet printing	Mimaki 3DUJ-553	SW-100	0.032

Fig. 2. Example of the 3D printed model, #7 made out of apricot kernel flour.

The comparative analysis of the 3D-printed copies fidelity was performed using Δs distances between the reference 'Artec surface model' and each of the surfaces obtained from the scanning of the respective physical copies. For the analysis, the system Geomagic Studio was used. Initial comparison consisted of minimization of the distances between two digital surfaces based on 2000 points. Further, detailed statistical analysis of the preliminary data was performed using CAD system PowerShape and Statistica software.

3 Results and Discussion

Analysis of Δs distances between the reference model and the respective scanned surfaces was somewhat limited by the application of a single fixation during the scanning of copies. As a result, the measuring points were collected from the upper side of the object only. The reference model, in turn, was a closed 3D surface, since the scanning procedure was performed from sides and the top with a further connection of the obtained points to one digital model. It caused somewhat overestimated values of Δs in some areas of the surface. In order to minimize overestimation, maximal distance between two compared surfaces was limited down to 5 mm. This limitation excluded from the statistical calculations some 24.7% of the surface at average, with maximal and minimal percentage 58% and 12%, respectively.

3.1 Results of Preliminary Comparison

The results of comparison between the reference digital surface and each of the scanned models were presented visually in form of a colored deviation map. Calculated maximal values of Δs distances above and below the reference surface were denoted as positive and negative, respectively. The average, apart from the overall one, was calculated additionally for negative and positive values. These results are shown in Table 2.

Table 2. Calculated distances Δs (mm) between the reference digital surface and scanned copies.

Copy no.	Maximum Positive	Maximum Negative	Average Overall	Average Positive	Average Negative	Standard deviation
#1	1.443	−1.894	−0.012	0.259	−0.193	0.273
#2	5.000	−2.272	0.044	0.165	−0.082	0.349
#3	2.445	−4.673	−0.015	0.069	−0.088	0.151
#4	2.494	−3.923	−0.053	0.093	−0.130	0.189
#5	–	–	–	–	–	–
#6	1.529	−2.240	−0.016	0.085	−0.096	0.116
#7	2.891	−3.675	−0.031	0.251	−0.232	0.299
#8	0.748	−2.860	−0.002	0.057	−0.061	0.075
#9	0.564	−0.696	−0.009	0.065	−0.073	0.086
#10	1.196	−2.971	−0.016	0.070	−0.078	0.094
#11	2.025	−3.570	−0.012	0.054	−0.067	0.089
#12	1.577	−3.637	−0.010	0.071	−0.088	0.105

Figures 3 and 4 present examples of the colored maps of Δs distances between the reference digital surface and scanned surfaces of copies #1 and #9, respectively. Large values of Δs distances above ± 1 mm are seen in Fig. 3, while only minor areas lay more than 0.2 mm above or below the reference surface.

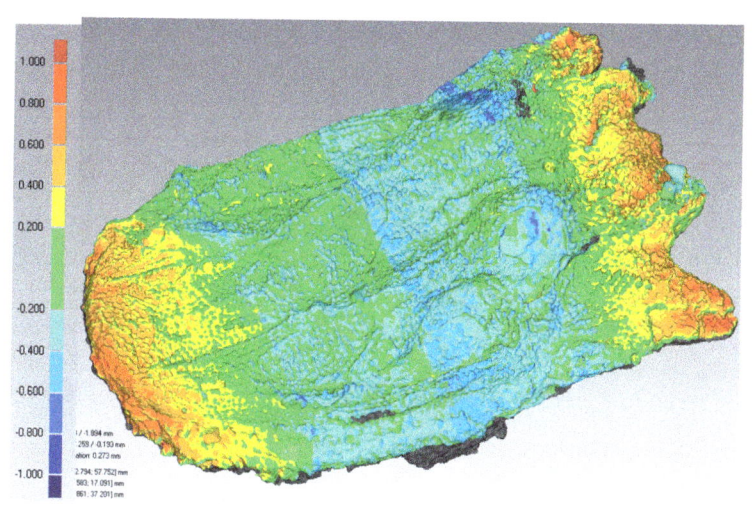

Fig. 3. Example of the colored map of Δs distances between reference digital surface and scanning results of copy #1.

From the visual analysis of the collected colored maps of deviations, copies #1, #4, #6, and #7 were found the most inaccurate. Especially copy #1 shown in Fig. 3 exhibited large areas inaccurately represented by the 3D-printing process.

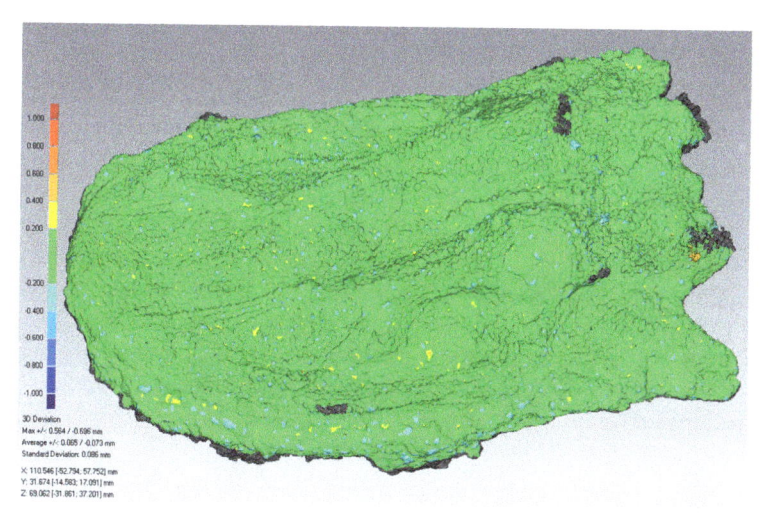

Fig. 4. Δs deviations between reference digital surface and scanning results of copy #9.

Based on the average distances from the reference model, it was found useful to plot a graph of the obtained standard deviations versus average distances. This is presented in Fig. 5.

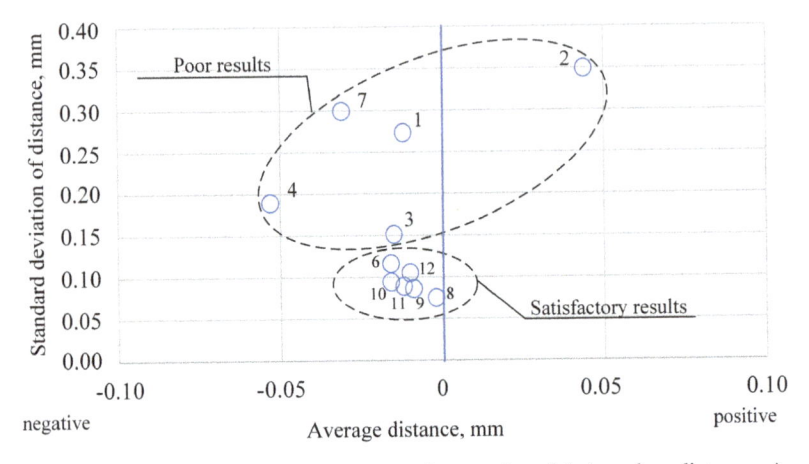

Fig. 5. Graphical analysis of the accuracy of scanned models based on distances Δs.

Model #2 exhibits the largest standard deviation combined with the largest average distance from the reference surface, hence, it can be considered the less accurate representation of the object. On the other hand, model #8 shows the smallest respective values, and as such it could be considered the most accurate one. However, because of small standard deviations close to 0.10 mm, some other models can be considered satisfactory and undergo further detailed analysis. The ones grouped as 'poor results' may be recommended against the use for 3D-printing of the analyzed fossil skull.

3.2 Statistical Analysis of the Results

Further statistical analysis of the obtained data was carried out using CAD system PowerShape. Due to the fact, that in different CAD systems different algorithms can be applied for calculations of the distances between digital surfaces, this approach provides reasonable ground for evaluation of the fidelity of the fossil skull copies.

Figure 6 presents graphically the results of the analysis of Δs distances distribution by means of the 'Box Whiskers' diagrams. The boxplot of multiple variables does not contain outliers and extremes because they bear no useful information. Their real values apparently do not fit the algorithm. Results in Fig. 6 allow for including copies #3 and #6 to the group of satisfactory results. In fact, this analysis demonstrates the complexity of the issue, so that decision on the feasibility of an AM method should not be based on a single parameter.

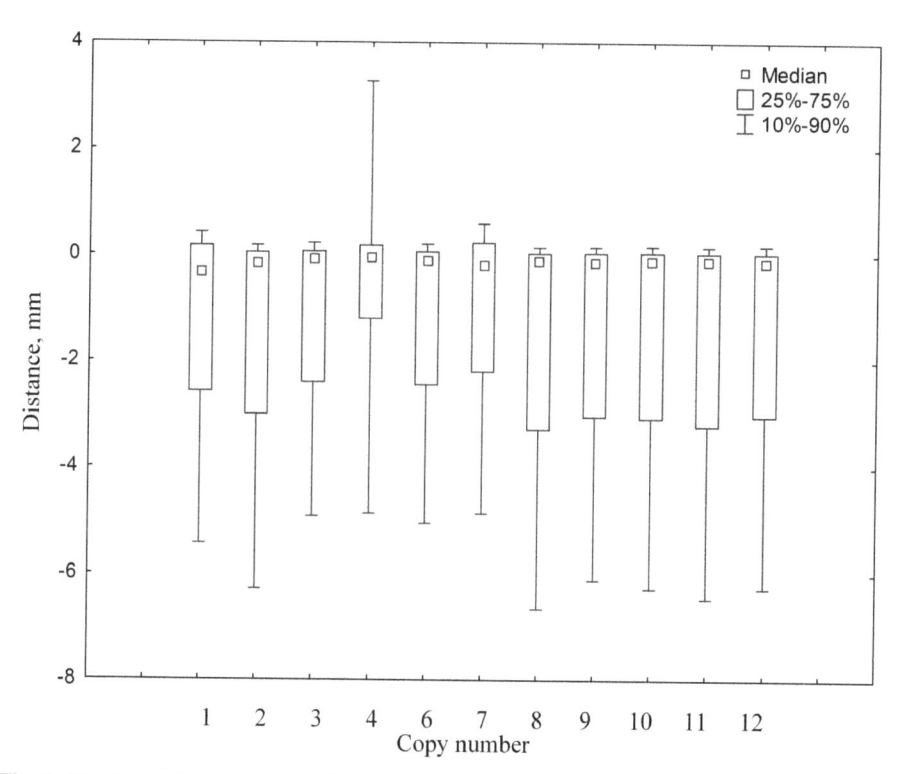

Fig. 6. Results of the statistical analysis of distances between the reference surface and scanned copies.

Thus, one more comparative analysis was performed. Namely, the scanned surfaces of the 3D-printed copies were compared to each other in couples in order to point out the surfaces closest to each other. Graphical presentation of the obtained standard deviations versus average distances between surfaces of each two copies is shown in Fig. 7.

Based on the results from Fig. 7, it was possible to arrange the copies in the rating list. The main criterion was the square root distance $\sigma\{\Delta s\}$ between the digitized surfaces of the respective copies. First, each couple was ascribed a number according to the value of $\sigma\{\Delta s\}$, and then the sum of these numbers determined the rating of each model. As a result, the copies were put in the following row from the closest one down to the one most distanced from others: #11 → #12 → #9 → #6 → #10 → #3 → #8 → #4 → #7 → #1.

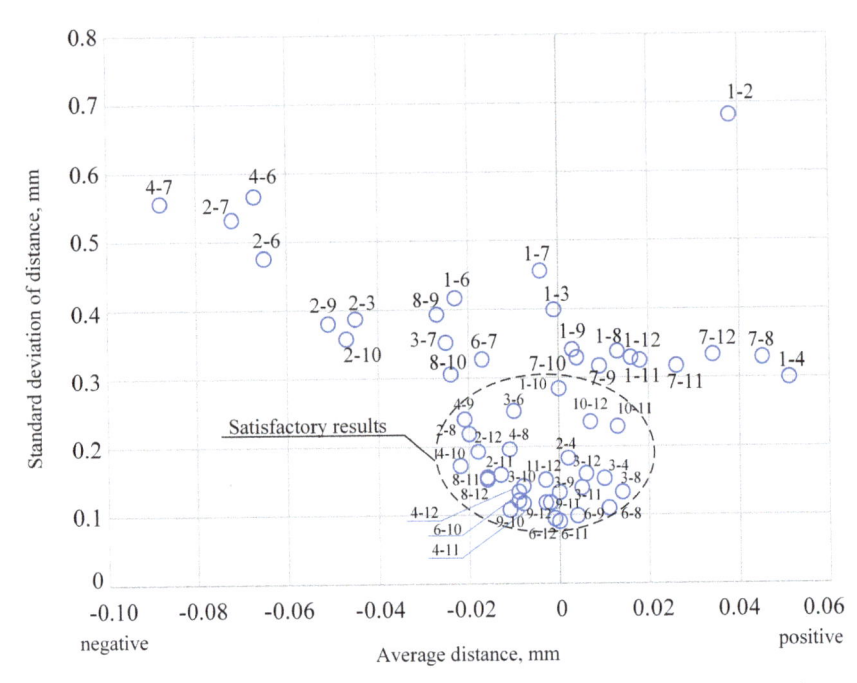

Fig. 7. Standard deviations versus average distances between the reference surface and scanned copies.

3.3 Detailed Statistics for Chosen Copies

From the abovementioned rating list, four of the first copies were chosen for the detailed statistical analysis using the Statistica software. Distances Δs between the reference digital surface and each scanned copy were taken from the results obtained from Autodesk PowerShape analysis. Figure 8 presents histograms and main statistical characteristics of the respective copies.

It is noteworthy that the obtained histograms are quite similar to one another, except for copy #6 that exhibits slightly different distribution of the distances Δs. For the square root distance $\sigma\{\Delta s\}$, for copies #6, #9, #12, and #11 the dispersion is rather narrow, from 0.0134 to 0.0159 mm. At the same time, median $Me\{\Delta s\}$ changes between -0.128 and -0.0979 mm, which can be considered a narrow range. These results confirm that the printed copies are close to each other in terms of surface reproduction fidelity.

Hence, all these four methods can be recommended as feasible for additive manufacture of the copies of a fossil skull similar to that of *Madygenerpeton* with an accuracy of ± 0.4 mm over the most of its surface. Reference to the abovementioned rating can be helpful, and other factors can be considered, such as price, availability, time, etc.

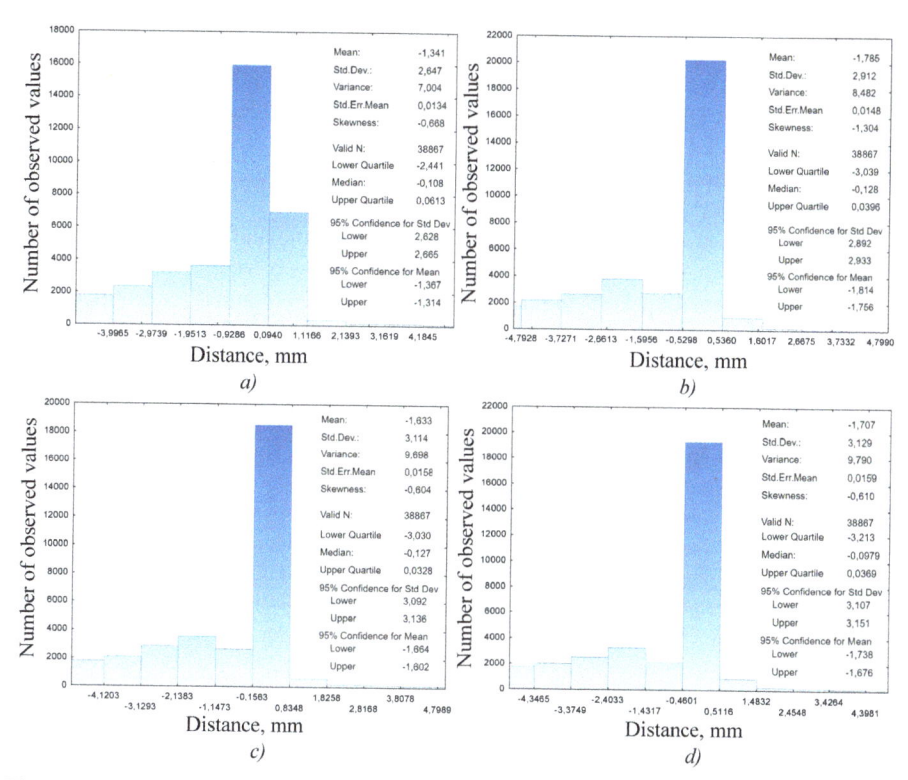

Fig. 8. Statistical parameters and histograms of distribution of distances Δs between the reference surface and scanned copies: a) #6, b) #9, c) #12, d) #11.

It can be presumed that a more accurate scanning method would provide better digital surfaces for the comparative analysis and, hence, would give more unequivocal recommendations. However, the optical surface properties of different copies, related to the surface texture, material characteristics, etc., make it impossible to find a single scanning method of the highest accuracy for all the printed copies.

Further research will cover a wider spectrum of the AM technologies able to produce reasonably accurate copies of the *Madygenerpeton* fossil skull. Additionally, it is interesting to consider different methods of scanning in order to find statistical background for comparison of the accuracy of the 3D-printed objects.

4 Conclusions

From the performed analysis of Δs distances between the reference model and the respective scanned surfaces of 3D-printed copies of *Madygenerpeton* fossil skull, the following conclusions can be derived.

First of all, there are certain limitations of the evaluation of the copies' accuracy due to different optical properties of the surfaces obtained from different AM methods. The same scanning method may provide highly accurate results for one 3D-printed copy, but not so accurate for another.

Next, the analysis based on Δs distances and on the standard deviation of their distribution throughout the scanned surface provided reliable results. From this analysis, a group of the copies of poor fidelity was identified. Further detailed statistics proved that the models identified as satisfactory accurate had similar distribution histograms, and based on the criterion of the square root distance $\sigma\{\Delta s\}$ between the digitized surfaces, the rating list was created.

And finally, it was demonstrated, that the proposed methodology was helpful in evaluation of the fidelity of additively manufactured copies of such a unique object as a *Madygenerpeton* fossil skull. Similar methodology can be applied for any paleontological object or a unique artifact where some individual features have to be copied and reproduced with high fidelity.

Acknowledgements. Results presented in this paper were achieved in the frame of the ESF-funded young researcher group "G.O.D.S." (Geoscientific Objects Digitization Standards) at the TU Bergakademie Freiberg. This paper has been supported by the Kazan Federal University Strategic Academic Leadership Program.

Authors express their gratitude to the colleagues who helped with the acquisition of 3D data, in particular Daniel Eger Passos and Sascha Schmidt (Freiberg), Maik Jähne, Henrik Alhers and Thomas Reuter (Dresden), Kristin Mahlow and Tom Cvjetkovic (Berlin), Tomasz Szymański and Robert Długoszewski (Mitutoyo Polska, Wrocław).

References

1. Zhang, X., Fan, W., Liu, T.: Fused deposition modeling 3D printing of polyamide-based composites and its applications. Compos. Commun. **21**, 100413 (2020)
2. Woźniak, J., Budzik, G., Przeszłowski, Ł., Chudy-Laskowska, K.: Directions of the development of the 3D printing industry as exemplified by the polish market. Manag. Prod. Eng. Rev. **12**(2), 98–106 (2021)
3. Gardan, J.: Additive manufacturing technologies: state of the art and trends. Int. J. Prod. Res. **54**(10), 3118–3132 (2016)
4. Elhoone, H., Zhang, T., Anwar, M., Desai, S.: Cyber-based design for additive manufacturing using artificial neural networks for Industry 4.0. Int. J. Prod. Res. **58**(9), 2841–2861 (2020)
5. Wang, Y., Lin, Y., Zhong, R.Y., Xu, X.: IoT-enabled cloud-based additive manufacturing platform to support rapid product development. Int. J. Prod. Res. **57**(12), 3975–3991 (2019)
6. Rosienkiewicz, M., Gabka, J., Helman, J., Kowalski, A., Susz, S.: Additive manufacturing technologies cost calculation as a crucial factor in industry 4.0. In: Hamrol A., Ciszak O., Legutko S., Jurczyk M. (eds.) Advances in Manufacturing. Lecture Notes in Mechanical Engineering, pp. 171–183. Springer, Cham (2018). https://doi.org/10.1007/978-3-319-68619-6_17
7. Tsirogiannis, E., Vosniakos, G.: Redesign and topology optimization of an industrial robot link for additive manufacturing. Facta Univ. Ser. Mech. Eng. **17**(3), 415–424 (2019)
8. Singh, R., et al.: Powder bed fusion process in additive manufacturing: an overview. In: Materials Today: Proceedings 26 (Part 2), pp. 3058–3070 (2020)
9. Schneck, M., Horn, M., Schmitt, M., Seidel, C., Schlick, G., Reinhart, G.: Review on additive hybrid- and multi-material-manufacturing of metals by powder bed fusion: state of technology and development potential. Prog. Addit. Manuf. **6**(4), 881–894 (2021)
10. Gibson, I., Rosen, D., Stucker, B., Khorasani, M.: Additive Manufacturing Technologies. 3rd edn. Springer, Cham (2021). https://doi.org/10.1007/978-3-030-56127-7

11. Gao, W., et al.: The status, challenges, and future of additive manufacturing in engineering. Comput. Aided Des. **69**, 65–89 (2015)
12. Di Angelo, L., Di Stefano, P., Guardiani, E.: Search for the optimal build direction in additive manufacturing technologies: a review. J. Manuf. Mater. Process. **4**(3), 71 (2020)
13. Dalpadulo, E., Pini, F., Leali, F.: Assessment of design for additive manufacturing based on CAD platforms. In: Rizzi, C., Andrisano, A.O., Leali, F., Gherardini, F., Pini, F., Vergnano, A. (eds.) ADM 2019. LNME, pp. 970–981. Springer, Cham (2020). https://doi.org/10.1007/978-3-030-31154-4_83
14. Zhang, Y., Xu, Y., Bernard, A.: A new decision support method for the selection of RP process: knowledge value measuring. Int. J. Comput. Integr. Manuf. **27**(8), 747–758 (2014)
15. Razavykia, A., Brusa, E., Delprete, C., Yavari, R.: An overview of additive manufacturing technologies — a review to technical synthesis in numerical study of selective laser melting. Materials **13**(17), 3895 (2020)
16. Di Angelo, L., Di Stefano, P., Marzola, A.: Surface quality prediction in FDM additive manufacturing. Int. J. Adv. Manuf. Technol. **93**(9–12), 3655–3662 (2017)
17. Chen, H., Zhao, Y.F.: Process parameters optimization for improving surface quality and manufacturing accuracy of binder jetting additive manufacturing process. Rapid Prototyping J. **22**(3), 527–538 (2016)
18. Li, N., et al.: Progress in additive manufacturing on new materials: a review. J. Mater. Sci. Technol. **35**, 242–269 (2019)
19. Schoch, R.R., Voigt, S., Buchwitz, M.: A chroniosuchid from the Triassic of Kyrgyzstan and analysis of chroniosuchian relationships. Zool. J. Linn. Soc. **160**, 515–530 (2010)
20. Voigt, S., et al.: Triassic life in an inland lake basin of the warm-temperate biome – the Madygen Lagerstätte (southwest Kyrgyzstan, Central Asia). In: Fraser, N.C., Sues, H.D. (eds.) Terrestrial conservation Lagerstätten. Windows into the evolution of life on land, pp. 65–104. Dunedin, Edinburgh, London (2017)
21. Buchwitz, M., Witzmann, F., Voigt, S., Golubev, V.: Osteoderm microstructure indicates the presence of a crocodylian-like trunk bracing system in a group of armoured basal tetrapods. Acta Zoologica **93**(3), 260–280 (2012)
22. Garashchenko, Y., Kogan, I., Rucki, M. Analysis of 3D triangulated models of Madygenerpeton pustulatum fossil skull. In: Euspen's 21st International Conference & Exhibition, pp. 89–90. Copenhagen, DK, June 2021
23. Kogan, I., Rucki, M., Jähne, M., Eger Passos, D., Cvjetkovic, T., Schmidt, S.: One head, many approaches – comparing 3D models of a fossil skull. In: Luhmann, T., Schumacher, C. (eds.), Photogrammetrie – Laserscanning – Optische 3D-Messtechnik: Beiträge der Oldenburger 3D-Tage 2020, pp. 22–31. Wichmann Verlag, Berlin (2020)

Analysis of the Direction-Dependent Point Identification Accuracy in CMM Measurement

Tomasz Mazur[1], Miroslaw Rucki[1(✉)], Michal Jakubowicz[2], and Lenka Cepova[3]

[1] Faculty of Mechanical Engineering, Kazimierz Pulaski University of Technology and Humanities in Radom, ul. Stasieckiego 54, 26-600 Radom, Poland
m.rucki@uthrad.pl
[2] Division of Metrology and Measurement Systems, Poznan University of Technology, Piotrowo Street 3, 60965 Poznan, Poland
[3] Department of Machining, Faculty of Mechanical Engineering, VSB-Technical University of Ostrava, Assembly and Engineering Metrology, 17. listopadu 2172/15, 70800 Ostrava, Poruba, Czech Republic

Abstract. The paper presents results of investigations on the direction-dependent accuracy of the point identification during contact probe measurement with a Coordinate Measuring Machine (CMM). Considering the contact point identified by orthogonal to the surface probe movement, transformation of coordinates was made in order to calculate the displacement of the measured point. As a result, positioning accuracy was estimated in three axes. The experiments demonstrated strong dependence of the displacement on the declination angle. Moreover, it was found that the directional surface texture which provided different roughness in perpendicular directions, had impact on the positioning accuracy.

Keywords: Coordinate measuring machine · Accuracy · Probing point

1 Introduction

Coordinate measuring machines (CMMs) perform a vast range of complex measurement tasks, from the simple length measurement to the verification of complex geometric tolerance including free-form surfaces [1]. CMMs are widely used for three-dimensional industrial measurement of workpieces using tactile probing systems. It is widely recognized that contacting CMM is rather favorable because it provides higher accuracy than the non-contacting one [2].

Usually, a stylus used for the contact measurement is equipped at the end with a ball tip to maintain the same form in all the directions. In such a complex system as CMM, "component errors overlap each other, determining the error vector for every point in the measuring volume" [3]. In this context, stylus tip has certain effect on the measurement results [4], and the point of contact between the ball tip and the surface of a measured object is of crucial importance in terms of measurement accuracy [5].

In order to increase accuracy of CMM measurement, many efforts are undertaken in various directions. Some researchers propose a procedure for selecting the most adequate probe orientations for measurement of parts with several elements being inspected

[6]. Others established the theoretical contact equation between the probe tip and the measured surface of the gear to evaluate the uncertainty of the gear measurement [7]. In the case of accurate measurement of aspheric surfaces, novel transformation algorithm for compensation of the error caused by the inaccurate radius of the probe tip, as well as the pre-travel errors compensation was presented [2]. This paper describes a radius compensation of the probe tip together with compensation of a slipping displacement from the predicted contact point for each measured point.

There can be found also a proposition of a new method of the trigger probe error compensation based on the Abbé measurement principle [8]. The authors proposed to measure the probe error directly, and thus to obtain the calibration error of each performance parameter of the probe, which made possible the error compensation of the measurement value in the measurement process.

The knowledge of CMM principles, and basic rules of operation is very important when assessing the measurement accuracy [9]. Considering importance of the single point uncertainty [10], the present study is focused on the difference between the real contact point and its identification by the CMM in the case when the movement of the probe ball tip is not perpendicular to the measured surface.

2 Materials and Methods

The experimental measurements were performed using a Coordinate Measuring Machine CNC Mitutoyo Crysta-Apex C7106 with the measurement head Renishaw PH10MQ. Its maximum permissible error according to ISO 10360-2 can be expressed as $MPE_E = 1.7 + 0.3\ L/100$ [μm]. The scanning probe SP-25M was used in measurement, with a ball tip of 2 mm diameter and turning arm length 173.35 mm.

The sample surface used in the experiments was made out of steel finished by face milling. In the area of planned contact with the CMM probe tip, the surface roughness was measured in the respective directions along probe tip movement denoted as $\alpha = 0°$ and perpendicular to it at $\alpha = 90°$. To perform the roughness measurement, Mitutoyo SJ-500P profilometer was used with diamond stylus. Evaluation length was 4 mm, with sampling length 0.8 mm, and respective cut-offs were $\lambda_c = 0.8$ mm and $\lambda_s = 0.0025$ mm. Ra parameter was measured 8 times in two perpendicular directions for $\alpha = 0°$ and $\alpha = 90°$, 4 times in each direction. It should be noted that substantial differences were found between Ra values obtained in these two directions, as it is shown in Table 1.

Table 1. Roughness of the measured surfaces.

Direction	Steel surface	α
Along probe movement	$Ra = 0.8$ μm	0°
Perpendicular to probe movement	$Ra = 0.3$ μm	90°

Substantial differences in *Ra* parameters of the same surface measured in different directions can be explained by the directional texture caused by movement of the finishing tool.

The measurements were performed in two perpendicular directions, at $\alpha = 0°$ and $\alpha = 90°$, using the local coordinate system. It was defined by the main plane *x-y* corresponding with the upper surface of the measured sample, measured in the automatic mode from 8 probing points evenly distributed on the circle of 10 mm diameter. The center of this circle was used as an initial point of the coordinate system *xyz* and denoted *A* as a contact point for the experiments. It was decided that the direction of *x*-axis corresponded with direction of maximal roughness (along probe movement in Table 1), while *y*-axis was determined by the direction of minimal roughness (perpendicular to probe movement in Table 1).

In the CMM's coordinate system, position of the point *A* was determined as $X = 410$, $Y = 410$ due to central position of the measured sample. The coordinate $Z = 184$ corresponded with the height of the sample.

3 Results and Discussion

3.1 Theoretical Calculations

The initial point of the further measurement assumes that the direction of movement *v* takes place along the *z*-axis perpendicularly to the sample surface. From this measurement illustrated in Fig. 1a, the reference point *A* was defined. Theoretically, when the *v* movement direction is not perpendicular, correction can be introduced. There is an option in the CMM software to rotate local coordinate system by angle β around the *x*-axis and to determine the position of reference point *A* from the real point *A'*. This geometrical operation is shown in Fig. 1b.

From the Fig. 1b it is seen that when the angle $\beta \neq 0°$, probe tip movement along new axis *z'* requires correction of the contact point position along new axis *y'*. The correction value is $(+R \times \sin \beta)$, where *R* is the radius of the ball tip. In addition, to keep the same approaching distance, correction along the new *z'*-axis is required by the value $(-R \times \sin \beta \times \operatorname{tg} \beta)$. Due to this correction, after declination by different values of β different points on the ball tip surface got in contact with the measured surface exactly in the same point *A*. Moreover, approaching distance was similar in all the measurements.

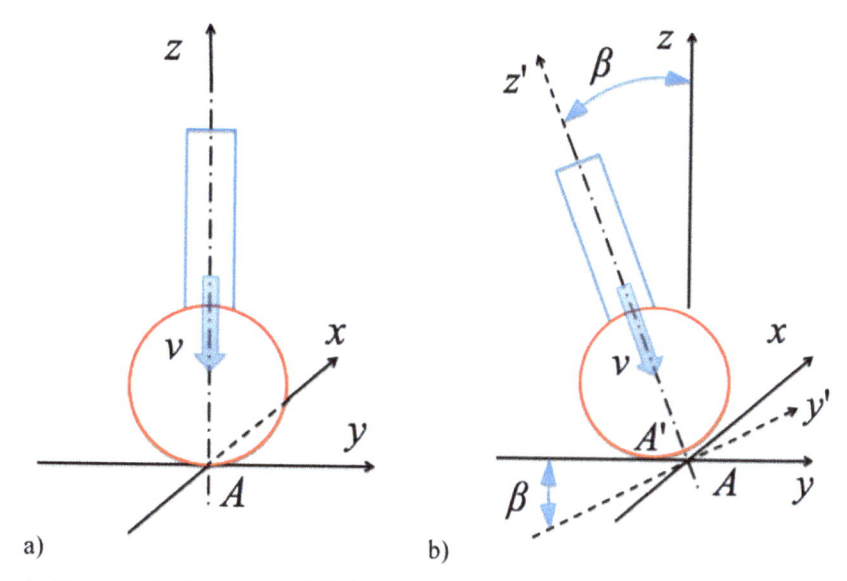

Analysis of the Direction-Dependent Point Identification Accuracy 51

Fig. 1. Contact point between the ball tip and the measured surface: a) obtained from the probe tip movement along the z-axis, defined as a reference point A, b) corrected from the real contact point A' through the rotation of the local coordinate system by angle β.

Comparison of the measurement results after declination of the probe with the reference results obtained for orthogonal direction of movement, it was necessary to work out the relevant equations. These equations were t transform coordinates of the points in x' y' z' system to the xyz system after subsequent rotations around z-axis by the angle $\alpha = 0°$ or $\alpha = 90°$, and then around x-axis by the angle β, as shown in Fig. 2.

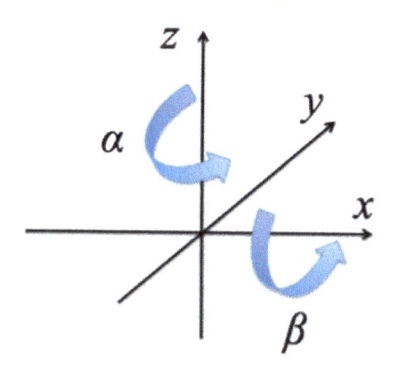

Fig. 2. Scheme of the rotations around z-axis by the angle $\alpha = 0°$ or $\alpha = 90°$, and around x-axis by the angle β.

The values of angle $\alpha = 0°$ and $\alpha = 90°$ correspond with directions where the maximal and minimal roughness Ra were detected, respectively.

The first partial transformation can be described by the matrix A_z while the second one by the matrix A_x, as follows:

$$A_z = \begin{bmatrix} \cos\alpha & \sin\alpha & 0 \\ -\sin\alpha & \cos\alpha & 0 \\ 0 & 0 & 1 \end{bmatrix} \quad (1)$$

$$A_x = \begin{bmatrix} 1 & 0 & 0 \\ 0 & \cos\beta & \sin\beta \\ 0 & -\sin\beta & \cos\beta \end{bmatrix} \quad (2)$$

When the transformation consists of two subsequent rotations described above, it can be written in form of Eq. (3), then all the coordinates of a point $[x', y', z']$ can be transformed to the xyz coordinate system using the Eq. (4).

$$A_{zx} = A_x \cdot A_z \quad (3)$$

$$\begin{bmatrix} x \\ y \\ z \end{bmatrix} = [A_x \cdot A_z]^T \begin{bmatrix} x' \\ y' \\ z' \end{bmatrix} \quad (4)$$

From the Eq. (4), the final equations for x, y, and z can be derived, as follows:

$$x = x' \cos\alpha - y' \sin\alpha \cos\beta + z' \sin\alpha \sin\beta \quad (5)$$

$$y = x' \sin\alpha + y' \cos\alpha \cos\beta - z' \cos\alpha \sin\beta \quad (6)$$

$$z = y' \sin\beta + z' \cos\beta \quad (7)$$

In addition, to assess the positioning accuracy, the parameter AP was considered as it was defined in the standard PN-ISO 9283 [11], which is very useful for the industrial robots [12]. It was calculated from 25 repetitions of each experiment in the same conditions. The following formula was applied:

$$AP = \sqrt{AP_x^2 + AP_y^2 + AP_z^2} = \sqrt{\left(\frac{1}{25}\sum_{i-1}^{25} x_i - x_c\right)^2 + \left(\frac{1}{25}\sum_{i-1}^{25} y_i - y_c\right)^2 + \left(\frac{1}{25}\sum_{i-1}^{25} z_i - z_c\right)^2} \quad (8)$$

where x_c, y_c, z_c are coordinates of the reference point A. Since these coordinates are different for $\alpha = 0°$ and $\alpha = 90°$, two sets of the results will be obtained.

3.2 Measurements for Different Angles B

The measurements were repeated for different angles β, starting from reference point at $\beta = 0°$, and continuing with 2°, 4°, and 8°. Each measurement was repeated 25 times. Example of the averages and standard deviations of points A and A' is shown in the Table 2.

Analysis of the Direction-Dependent Point Identification Accuracy   53

Table 2. Averages and standard deviations of the coordinates for 25 repetitions at $\alpha = 0°$.

	$\beta = 0°$ (reference)	$\beta = 2°$	$\beta = 4°$	$\beta = 8°$
x'	−0.0220	−0.0220	−0.0220	−0.0221
y'	−0.0241	0.0140	0.0526	0.1286
z'	0.0001	0.0002	−0.0012	−0.0078
$\pm 3s_{x'}$	2.12×10^{-17}	2.12×10^{-17}	2.12×10^{-17}	0.000831
$\pm 3s_{y'}$	0.000831	0.001364	0.00147	0.002722
$\pm 3s_{z'}$	0.000831	0.001308	0.001122	0.001122
X	−0.0220	−0.0220	−0.0220	−0.0221
Y	−0.0241	0.0140	0.0526	0.1285
Z	0.0001	0.0007	0.0025	0.0101

In order to minimize impact of the probe pretravel [13–15], in all the measurements, the procedure was started from the point where tip ball center was distanced 5 mm from the reference point A measured along the movement path. This value corresponded with the reference point at $\beta = 0°$ and ensured pretravel value of 4 mm. The same pretravel value was kept for all other directions, where $\beta \neq 0°$, which required recalculation of the probe ball tip center start point but ensured the same pretravel and the same contact point.

The results of the measurements are presented graphically in Fig. 3 in three planes, x'-y', x'-z', and y'-z', respectively. In the graphs, usually results are grouped together for a particular declination angle β, but in some cases these groups overlapped without distinguishable differences between them.

The series of the measurements exhibited very close results for each value of β angle. It can be noted that for $\beta = 0°$, i.e. for the reference point A, results were close to the point (0, 0, 0) in the planes x'-y' and y'-z', namely (−0.022, −0.024, 0.001) for $\alpha = 0°$ and (−0,024; 0,022, 0,002) for $\alpha = 90°$. The next series collected for $\beta = 2°$ is grouped around a point with larger absolute values of coordinates, demonstrating that the larger β is, the longer is distance from A' to the reference point A. It should be noted that the points lay along the direction of the probe movement oy' in the plane x'-y' around A point, and for each increasing value of angle β they are displaced farther from A. No distinguishable displacement was found along ox' axis, perpendicular to the movement direction.

Corrected results for the respective reference points A are shown in Fig. 4.

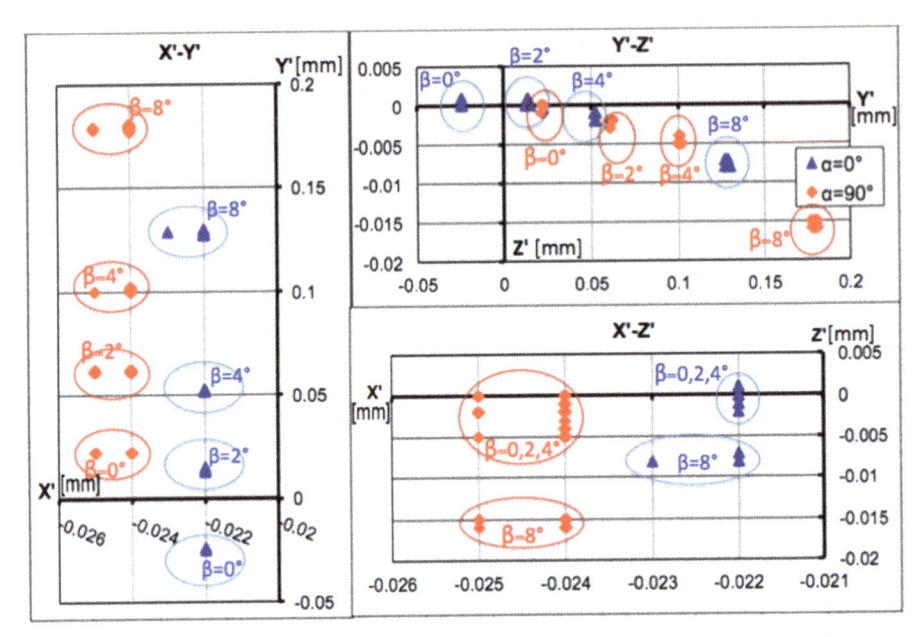

Fig. 3. Uncorrected results of the measurements in the coordinate system $x'y'z'$.

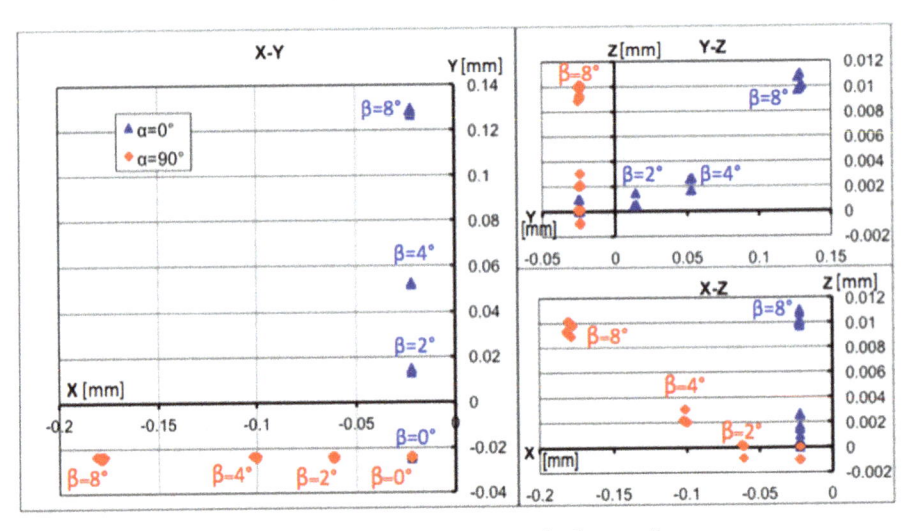

Fig. 4. Corrected results of the measurements in the coordinate system xyz.

3.3 Positioning Accuracy

Since position errors associated to the CMM axis displacement are important [16, 17], of higher interest was the calculation of the positioning accuracy AP from the Eq. (8). In the Table 3, there are collected results for each axis, AP_x, AP_y, AP_z, calculated for $\alpha = 0°$ and $\alpha = 90°$, respectively.

Table 3. Analysis of positioning accuracy.

	$\alpha = 0°$			$\alpha = 90°$		
	$\beta = 2°$	$\beta = 4°$	$\beta = 8°$	$\beta = 2°$	$\beta = 4°$	$\beta = 8°$
AP_x	0	0	−0.0001	−0.0394	−0.0789	−0.1577
AP_y	0.0381	0.0767	0.1526	0	0.0002	$\times 10^{-4}$
AP_z	0.0006	0.0024	0.01	0.0003	0.0024	0.01
AP	0.038105	0.076738	0.152927	0.039401	0.078937	0.158017

From the Table 3 and the graph x-y in Fig. 4 it can be noted that the respective coordinates are displaced along the probe movement both for $\alpha = 0°$ and $\alpha = 90°$. However, the displacements are not uniform, since higher roughness in direction $\alpha = 90°$ caused smaller values of displacement than that for $\alpha = 0°$. At small angles β, the corrected points are displaced from the reference point (0, 0, 0), but still lay in the plane x-y, i.e. on the measured surface. However, larger declination β caused increase of the calculated value z, moving the point away from its real position. It was calculated initially, that at $\beta = 30°$, AP_z can be as high as 0.13 mm.

It should be noted also that positioning accuracy in the direction perpendicular to the probe movement kept small values, while in other directions AP was increasing for larger angles β, indication the worsening of measurement accuracy. For both angles $\alpha = 0°$ and $\alpha = 90°$, AP values are similar at the same declinations β, respectively.

4 Conclusions

From the presented analysis, the following important conclusions can be drawn. First of all, clear dependence of the point identification on the probe ball tip movement direction was demonstrated. The higher declination angle is, the larger is the displacement of the identified point from its real position.

In terms of positioning accuracy, it was found that increase of declination angle β caused displacement also in z-axis, moving the identified point away from its real position on the measured surface. Noteworthy, AP values were similar at the same declinations β, irrespective on the actual movement angle $\alpha = 0°$ or $\alpha = 90°$, that determined machining-dependent directional roughness. It is important to investigate in further researches the impact of other materials and surface texture on the positioning accuracy.

Moreover, it is planned to perform additional measurements and calculations for other dimensions of a probe ball tip. To obtain better insight to the phenomena, it will be necessary also to widen the range of the angles β in further researches.

Acknowledgement. The paper was prepared in the frames of mutual academic training between Poznan University of Technology and Kazimierz Pulaski University of Technology and Humanities in Radom.

References

1. Petrò, S., Moroni, G.: A statistical point of view on the ISO 10360 series of standards for coordinate measuring systems verification. Measurement **172**, 108937 (2021). https://doi.org/10.1016/j.measurement.2020.108937
2. Ahn, H.K., Kang, H., Ghim, Y.-S., Yang, H.-S.: Touch probe tip compensation using a novel transformation algorithm for coordinate measurements of curved surfaces. Int. J. Precis. Eng. Manuf. **20**(2), 193–199 (2019). https://doi.org/10.1007/s12541-019-00076-2
3. Sładek, J.A.: Coordinate Metrology. Springer, Berlin (2016). https://doi.org/10.1007/978-3-662-48465-4_4
4. Zelinka, J., Čepová, L., Gapiński, B., Čep, R., Mizera, O., Hrubý, R.: The effect of a stylus tip on roundness deviation with different roughness. In: Diering, M., Wieczorowski, M., Brown, C.A. (eds.) MANUFACTURING 2019. LNME, pp. 147–157. Springer, Cham (2019). https://doi.org/10.1007/978-3-030-18682-1_12
5. Ito, S., Tsutsumi, D., Kamiya, K., Matsumoto, K., Kawasegi, N.: Measurement of form error of a probe tip ball for coordinate measuring machine (CMM) using a rotating reference sphere. Precis. Eng. **61**, 41–47 (2020). https://doi.org/10.1016/j.precisioneng.2019.09.017
6. Martínez-Pellitero, S., Barreiro, J., Cuesta, E., Fernández-Abia, A.I.: Knowledge base model for automatic probe orientation and configuration planning with CMMs. Robot. Comput. Integr. Manuf. **49**, 285–300 (2018). https://doi.org/10.1016/j.rcim.2017.08.012
7. Yin, P., Han, F., Wang, J., Lu, C.: Influence of module on measurement uncertainty of gear tooth profile deviation on gear measuring center. Measurement **182**, 109688 (2021). https://doi.org/10.1016/j.measurement.2021.109688
8. Ren, G., Qu, X., Chen, X.: Performance evaluation and compensation method of trigger probes in measurement based on the Abbé Principle. Sensors **20**(8), 2413 (2020). https://doi.org/10.3390/s20082413
9. Kopáčik, A., Erdélyi, J., Kyrinovič, P.: Coordinate Measuring Systems and Machines. In: Engineering Surveys for Industry. Springer, Cham (2020). https://doi.org/10.1007/978-3-030-48309-8_7
10. Müller, A.M., Hausotte, T.: Determination of the single point precision associated with tactile gear measurements in scanning mode. J. Sens. Sens. Syst. **9**, 61–70 (2020). https://doi.org/10.5194/jsss-9-61-2020
11. PN-ISO 9283:2003. Manipulating industrial robots – Performance criteria and related test methods
12. McGarry, L., Butterfield, J., Murphy, A.: Assessment of ISO Standardisation to identify an industrial robot's base frame. Robot. Comput. Integr. Manuf. **74**, 102275 (2022). https://doi.org/10.1016/j.rcim.2021.102275
13. Woźniak, A., Dobosz, M.: Metrological feasibilities of CMM touch trigger probes. Part I: 3D theoretical model of probe pretravel. Measurement **34**(4), 273–286 (2003). https://doi.org/10.1016/j.measurement.2003.05.001
14. Cai, Y.L., Cui, N.N., Mo, X., Yao, X.K., Sun, W.Q.: The pre-travel error study of electrical trigger probe in on-machine measurement. Key Eng. Mater. **693**, 1466–1473 (2016). https://doi.org/10.4028/www.scientific.net/KEM.693.1466
15. Li, Y., Zeng, L., Tang, K., Li, S.: A dynamic pre-travel error prediction model for the kinematic touch trigger probe. Measurement **146**, 689–704 (2019). https://doi.org/10.1016/j.measurement.2019.07.005
16. Franco, P., Jodar, J.: Theoretical analysis of straightness errors in coordinate measuring machines (CMM) with three linear axes. Int. J. Precis. Eng. Manuf. **22**(1), 63–72 (2020). https://doi.org/10.1007/s12541-019-00264-0
17. Mekid, S.: Metrology and Instrumentation: Practical Applications for Engineering and Manufacturing. Wiley and ASME Press, Hoboken (2022)

Non-destructive Testing of Metal-Polymer Laminates by Digital Shearography

Zuzanna Konieczna[✉], Frans Meijer, and Ewa Stachowska

Division of Metrology and Measurement Systems, Institute of Mechanical Technology, Poznan University of Technology, Piotrowo 3, 60-965 Poznań, Poland
zuzanna.konieczna@doctorate.put.poznan.pl,
ewa.stachowska@put.poznan.pl

Abstract. This paper presents non-destructive detection of defects in metal-polymer laminates using digital shearography. A shearography setup built from of the shelf components was used together with self-written data evaluation software. Testing was performed on a polymer strap glued to a metal one of the same size, with purposely made breaks in the glue layer. Connection defects, their location, size and shape could be detected using three different defect sizes; the smallest defect was 6 mm wide. Testing required 10 s of heating; the data evaluation needed less than a minute. The temperature of the samples was raised by about 2 °C on the metal side and 0.5 °C on the polymer side. Relaxation of the samples was also observed. The small rise of temperature and the fast return to a state of equilibrium allows for multiple testing of samples without causing damage.

Keywords: Digital shearography · NDT · Connection defects · Laminates

1 Introduction

Over the last decade composite materials became not only common, but almost ubiquitous. They are widely used in cars, aero-space, wind turbines and many more applications. The use of specific composite materials depends on their properties, including reliability. In the case of laminates, the reliability of the material depends on the connection quality between its components, often polymers and metals [1, 2]. There are many testing methods to examine the strength of that connection; unfortunately, in many cases it is unavoidable to destroy the object investigated. That makes it impossible to further test or reuse it.

With the development of many industries comes not only the need to develop new and better performing materials, but also the need for new methods to test them. It is very desirable to use Non-Destructive Testing (NDT) when looking for hidden (from sight) defects such as delamination, disbonding, fiber breakage and more. While investigating different testing methods it is common to use composites made with cuts in the woven fibers or polymers glued to metal plates with local gaps in the glue layer. The use of ultrasonic measurements, thermography, vibrometry and shearography is well established [3–6]. Shearography proves to be a fast method, it also allows to change loading techniques to whichever is most appropriate for a given material. The use of shearography for materials testing is well established, also for different composites. We used it to investigate the possibility to detect defects of bonding in layered materials, especially metal-polymer laminates [7–11]. The load used is usually pressure (up to 0.5 atm) or mechanical deformation (tens of μm) and the defect sizes investigated range from 15 mm to 40 mm in width or diameter [12–15]. We investigated the possibility of detecting smaller defects in bonding of metal-polymer laminates, using small and fast thermal loading.

2 Experimental

The goal of our research was to detect unknown and invisible defects of the glue layer connecting polymer and metal using shearography and thermal loading. We aimed to detect defects smaller than 15 mm, but at the same time we wanted our shearography setup to be as versatile as possible when it comes to defect or object size or structure. The modular character of the setup allows just that.

2.1 Digital Shearography Setup

Shearography is a non-destructive measuring method that provides information about the gradient of the surface deformation. This deformation is the effect of applying a load to the object under investigation. This load can be mechanical, thermal or pressure or vibration. In digital shearography an image of the surface of the object in unloaded as well as loaded state is registered by a solid-state sensor such as a CCD or CMOS sensor array. Each time the speckle pattern is imaged through a shearing device such as a Michelson interferometer, so the camera records an interferogram. The two shearograms are subtracted and a fringe pattern is obtained. The quantity measured is actually the first derivative of deformation that corresponds to an underlying defect [16–18].

Our shearography system is a self-built modular setup. It consists of opto-mechanical elements and a CMOS camera (1280 × 1024 pixels) from ThorLabs [19], a He-Ne laser (λ 632.8 nm) and a computer (see Fig. 1) with our own software. The He-Ne laser guarantees stable illumination. A cylindrical lens and a diffusing lens mounted on cage system construction rods allow us to illuminate samples of different sizes by changing the distances between the lenses and the mirror reflecting light into them. The imaging lens is also mounted on rods, so the object distance can be changed to alter the magnification. Also, the distances between the beamsplitter, the mirrors, the imaging lens and the camera can be changed in a similar way. It is therefore possible to easily adjust the setup for observation of samples from millimeters wide to tens of centimeters wide. While the design of the shearography system is consistent with known and established setups, building it from simple opto-mechanical elements allows us to adjust the system in an easy and fast way to adapt to the object under study. Building our own setup this way is also a cheaper solution and allows us to use our own software making the way the results are obtained more transparent.

To record images an in-house modification of the C++ program uEye provided by Imaging Development Systems GmbH was used. It allows for automatic registration of a series of images with a user defined time step between them. A program to subtract images, filter and present the results was written in "R" [20], an open-source programming language with very well developed statistical and visualization facilities. It allows to use the matrix containing the information about the pixel intensity provided by the camera software in a fast and efficient way. The R-packages "plot3D" and "EBImage" were used for the visualization of results. The size of the object in the visualization is given in pixels. The ratio between mm's on the object and pixels on the camera can easily be determined using a grid of known dimensions as object. The pixel intensity corresponds directly to the light intensity in the interferograms.

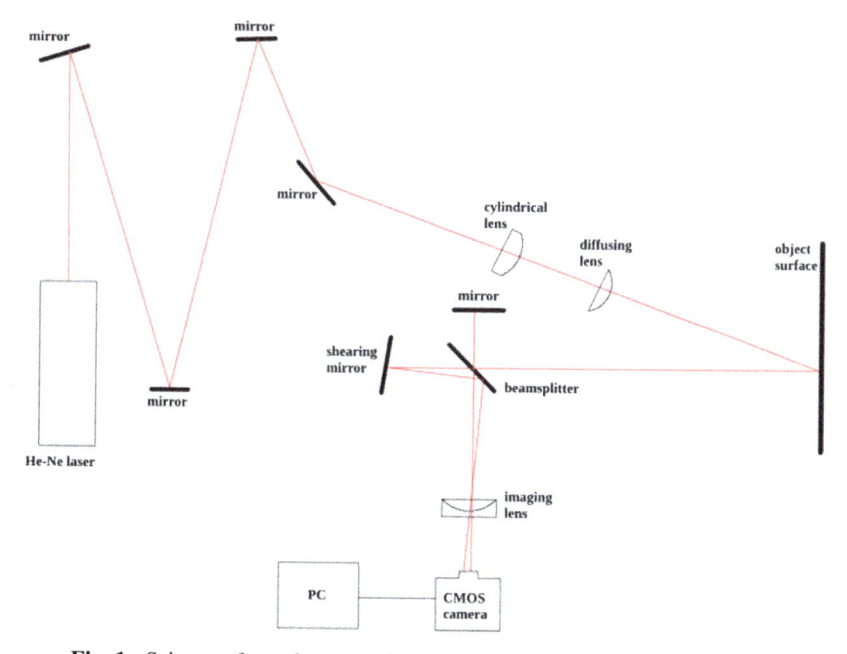

Fig. 1. Scheme of our shearography setup with a Michelson interferometer.

2.2 Sample Preparation

The samples used for the experiment were made from injection molding grade polyamide 6 Tarnamid T-27 and non-alloy quality steel DC04 glued together with epoxy Araldite 2011 as described in the paper by Nowak-Grzebyta et al. [21]. The size of the samples was 150 mm × 10 mm × 5 mm (length × width × height). Edge to edge defects of three different widths (resp. 18 mm, 12 mm and 6 mm) were created in the epoxy layer approximately in the middle of the samples (see Fig. 2). After a closer inspection under a microscope and through the polyamide layer, illuminated from the side of the sample, we noticed that the defect in the glue layer was not rectangular: near the edges of the samples the polymer part and the metal one were not fully glued together.

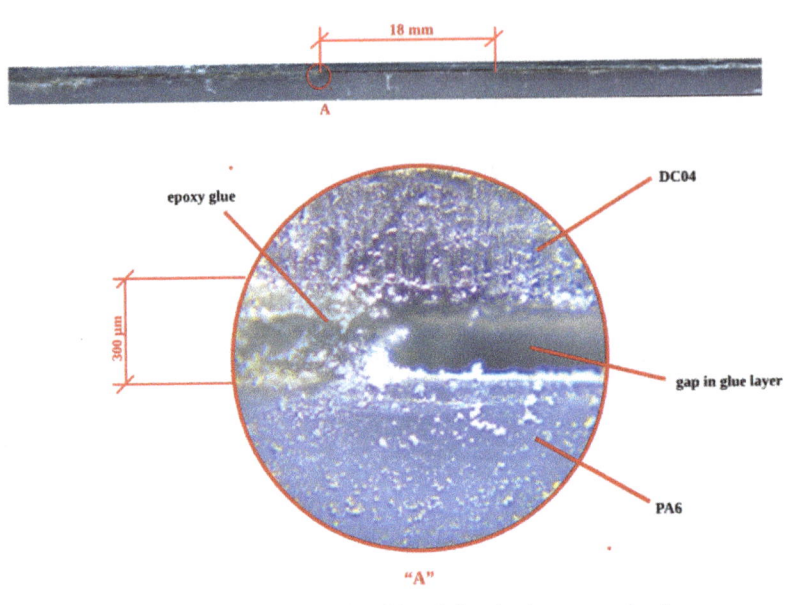

Fig. 2. Side view of a sample with a defect in the epoxy glue layer.

2.3 Sample Mounting and Loading

The mounting setup was placed on a pneumatic vibration-isolated optical table. The sample was mounted with clamps at each end. The polymer side was facing the interferometer (see Fig. 3). The whole shearography setup was mounted on the same table. The surface of the sample was coated with scanning spray from AESUB to obtain the roughness needed for the speckle pattern to appear. The spray does not affect the material beneath it and can be easily washed off. The area of the sample under observation was 71 mm × 10 mm (length x width), the mounting wasn't visible (see Fig. 4). For our experiment thermal loading was used. The sample was heated on the metal side using a 150 W infrared lamp through a mask allowing for even heating of the sample and preventing heating the rest of the setup. Heating lasted 10 s, raising the temperature of the steel surface by 1.5–2 °C and the polyamide surface by 0.5 °C (measured with a Termio 31 temperature recorder connected to a thermocouple). Image registration was done during both heating and cooling.

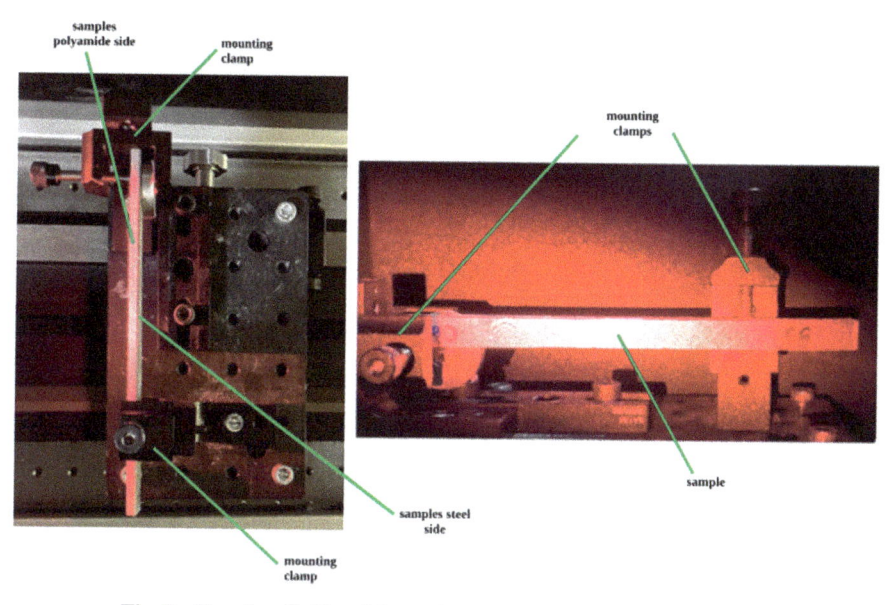

Fig. 3. Top view (left) and front view (right) of the sample mounting.

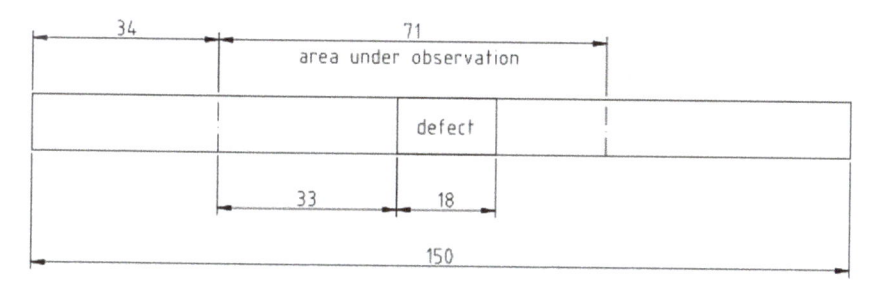

Fig. 4. Visible area of the sample.

3 Results and Discussion

3.1 Defect Detection

Using an image in the unloaded state and one captured after 10 s of heating all defects were detected as shown on the figures below (see Fig. 5, 6 and 7). In such a short period of heating the larger defect can be seen as an edge to edge defect; the smaller ones show deformation only closer to the edges, but are still detected. Also, the deformation (correlated with the pixel intensity and using the color scale in the visualization of the result) is larger for the wider defects and gets smaller and smaller with the reduction of the defect width. The not fully connected edges are also visible, proving that the optical resolution of our shearograph is at least 1 mm. The detected defects are consistent with defects detected using vibrometry [21] and the defect sizes are smaller than known from literature for composites testing using digital shearography [12–15].

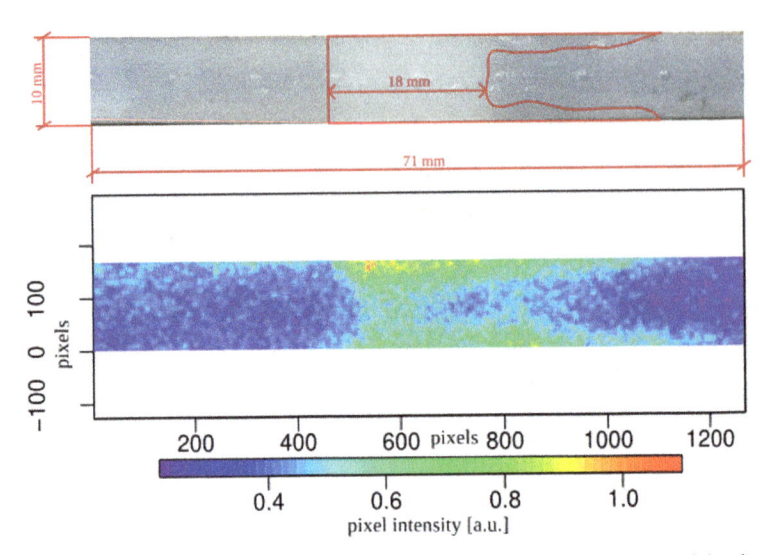

Fig. 5. Sample with an 18 mm defect: a photo with the defect indicated (upper) and the shearogram (lower); both images show the same area; 1mm is 18 pixels.

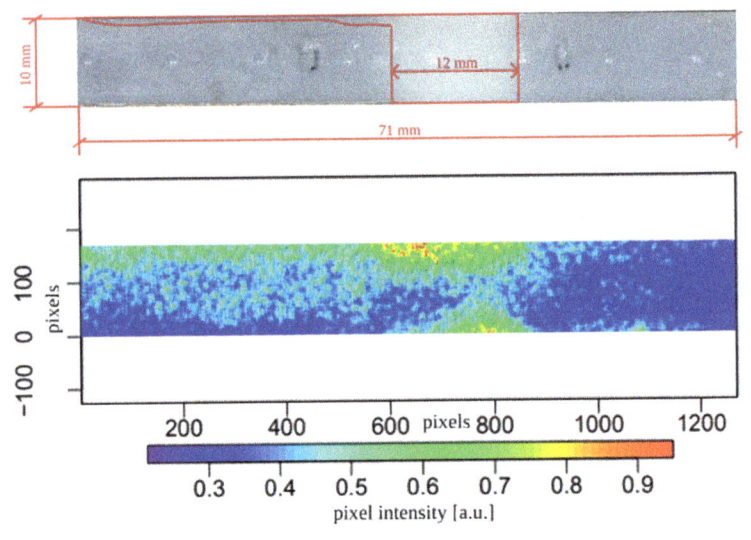

Fig. 6. Sample with a 12 mm defect: a photo with the defect indicated (upper) and the shearogram (lower); both images show the same area; 1 mm is 18 pixels.

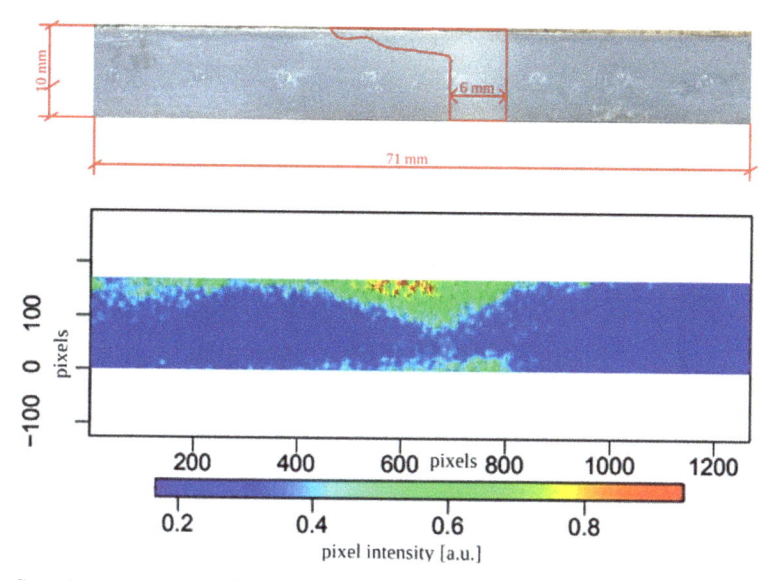

Fig. 7. Sample with a 6 mm defect: a photo with the defect indicated (upper) and the shearogram (lower) both images show the same area; 1 mm is 18 pixels.

3.2 Relaxation

Relaxation was also investigated, using three times longer heating time. The sample was heated for 30 s, then an image was registered every 2 s during 30 s cooling. Subtracting subsequent images (first and second, second and third and so on) made it possible to investigate the change of deformation in steps of two seconds; i.e. observing the relaxation of the sample (see Fig. 8 and 9). During the first ten seconds the changes are rapid, although slowing down. After 15 s the level of the unloaded setup is reached (the deformation observed without applying a load during measurement). The relaxation is fast, suggesting that not much load is put to the sample. The sample quickly returns to its normal state even after long heating. To detect defects only 10 s of heating is needed, so the samples are even less affected.

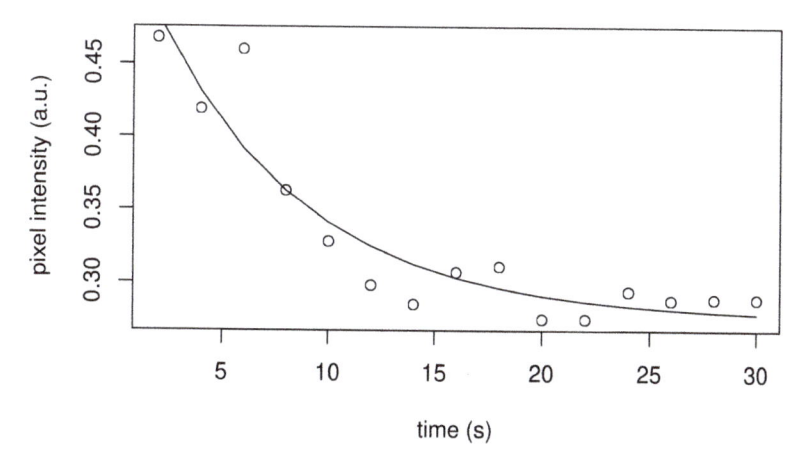

Fig. 8. Change of maximum pixel intensity during cooling.

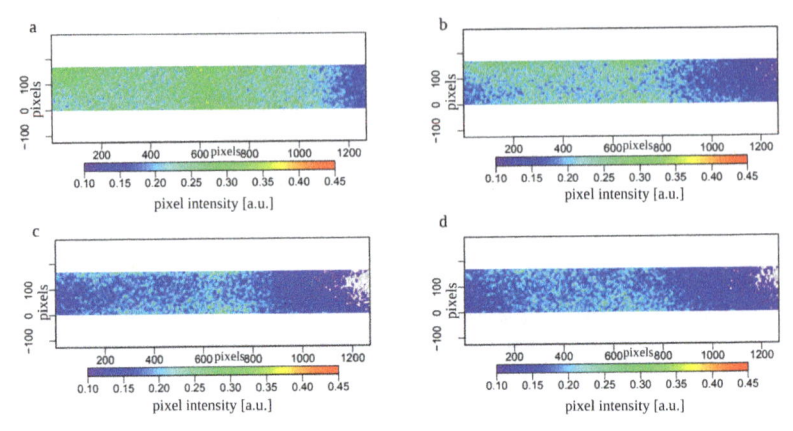

Fig. 9. Change of deformation of the polymer side during the third (a), seventh (b), eleventh (c) and twenty-first (d) second of cooling (the white colors represent areas where the pixel intensity is smaller than 0.10).

4 Conclusions

Even small defects change the way the laminate behaves under vibration [21], which can lead to possible construction failures. It is therefore important to detect such defects, also in components during production. Using a simple, inexpensive modular shearography setup it is possible to detect defects, their location, shape and size in metal-polymer laminates. It is also possible to observe the relaxation process of the material, which can important to predict long-term damage. The results are comparable with vibrometry, which is a much more complicated setup and is not suitable for relaxation observation. Considering the temperature rise and the relaxation time we can conclude that digital shearography is a useful non-destructive method to test for small defects in laminate structures. Using a measuring system constructed from of the shelf opto-mechanical parts, adaptation to different sample sizes is easy. It is also a fast testing method. It is therefore useful not only for laboratory measurements, but also as an in-line measuring system during production.

References

1. Venables, J.D.: Adhesion and durability of metal/polymer bonds. In: Mittal, K.L. (ed.) Adhesive Joints. Springer, Boston (1984). https://doi.org/10.1007/978-1-4613-2749-3_27
2. Grujicic, M., Sellappan, V., Omar, M.A., Seyr, N., Obieglo, A., Erdmann, M., et al.: An overview of the polymer-to-metal direct-adhesion hybrid technologies for load-bearing automotive components. J. Mater. Process. Technol. **197**(1–3), 363–373 (2008). https://doi.org/10.1016/j.jmatprotec.2007.06.058
3. Duchene, P., Chaki, S., Ayadi, A., Krawczak, P.: A review of non-destructive techniques used for mechanical damage assessment in polymer composites. J. Mater. Sci. **53**(11), 7915–7938 (2018). https://doi.org/10.1007/s10853-018-2045-6

4. Krasnoveikin, V.A., Konovalenko, I.S.: Development of the noncontact approach to testing the dynamic characteristics of carbon fiber reinforced polymer composites. In: 12th International Conference on Mechanics, Resource and Diagnostics of Materials and Structures (2018). https://doi.org/10.1063/1.5084486
5. Adebahr, W., Solodov, I., Rahammer, M., Gulnizkij, N., Kreutzbruck, M.: Local defect resonance for sensitive non-destructive testing. In: AIP Conference Proceedings, vol. 1706, p. 050005 (2016). https://doi.org/10.1063/1.4940504
6. Dehui, W., Fan, Y., Teng, W., Wenxiong, C.: A novel electromagnetic nondestructive testing method for carbon fiber reinforced polymer laminates based on power loss. Compos. Struct. **276** (2021). https://doi.org/10.1016/j.compstruct.2021.114421
7. Kadlec, M., Ruzek, R.: A comparison of laser shearography and c-scan for assessing a glass/Epoxy laminate impact damage. Appl. Compos. Mater. **19**, 1–15 (2011). https://doi.org/10.1007/s10443-011-9211-1
8. Kim, G., Hong, S., Jhang, K.Y., et al.: NDE of low-velocity impact damages in composite laminates using ESPI, digital shearography and ultrasound C-scan techniques. Int. J. Precis. Eng. Manuf. **13**, 869–876 (2012). https://doi.org/10.1007/s12541-012-0113-4
9. Théroux, L.D., Dumoulin, J., Maldague, X.: Active thermal shearography and infrared thermography applied to NDT of reinforced concrete structure by glued CFRP. In: EWSHM - 7th European Workshop on Structural Health Monitoring, IFFSTTAR Nantes, France (2014)
10. Angelis, G., Dati, E., Bernabei, M., Leccese, F.: Development on aerospace composite structures investigation using thermography and shearography in comparison to traditional NDT methods. In: 2nd IEEE International Workshop on Metrology for Aerospace, MetroAeroSpace 2015 – Proceedings, pp. 49–55 (2015). https://doi.org/10.1109/MetroAeroSpace.2015.7180625
11. Sun, J., Wang, Y., Gao, X., et al.: Dynamic measurement of first-order spatial derivatives of deformations by digital shearography. Instrum. Exp. Tech. **60**, 575–583 (2017). https://doi.org/10.1134/S0020441217040145
12. Zhao, Q., Dan, X., Sun, F., Wang, Y., Wu, S., Yang, L.: Digital shearography for NDT: phase measurement technique and recent developments. Appl. Sci. **8**(12), 2662 (2018). https://doi.org/10.3390/app8122662
13. Taillade, F., Quiertant, M., Benzarti, K., Aubagnac, C.: Non-destructive evaluation (NDE) of composites: using shearography to detect bond defects. In: Karbhari, V.M. (eds), Woodhead Publishing Series in Composites Science and Engineering, Non-Destructive Evaluation (NDE) of Polymer Matrix Composites 542–557e, Woodhead Publishing (2013). https://doi.org/10.1533/9780857093554.4.542
14. Georges, M., Srajbr, C., Menner, P., Koch, J., Dillenz, A.: Thermography and shearography inspection of composite hybrid sandwich structure made of CFRP and GFRP core and titanium skins. Proceedings. **2**(8), 484 (2018). https://doi.org/10.3390/ICEM18-05384
15. De Angelis, G., Meo, M., Almond, D.P., Pickering, S.G., Angioni, S.L.: A new technique to detect defect size and depth in composite structures using digital shearography and unconstrained optimization. NDT E Int. **45**, 91–96 (2012). https://doi.org/10.1016/j.ndteint.2011.07.007
16. Yang, L., Li, J.: Shearography. In: Ida, N., Meyendorf, N. (eds.) Handbook of Advanced Nondestructive Evaluation, pp. 383–419. Springer, Cham (2019). https://doi.org/10.1007/978-3-319-26553-7_3
17. Francis, D., Tatam, R.P., Groves, R.M.: Shearography technology and applications: a review. Meas. Sci. Technol. (2010). https://doi.org/10.1088/0957-0233/21/10/102001

18. Steinchen, W., Yang, L.: Digital Shearography: Theory and Application of Digital Speckle Pattern Shearing Interferometry. SPIE, Bellingham (2003)
19. ThorLabs Homepage. https://www.thorlabs.com/. Accessed 14 Oct 2021
20. The R Project for Statistical Computing. https://www.r-project.org/about.html. Accessed 10 Oct 2021
21. Nowak-Grzebyta, J., Meijer, F., Bula, K., Stachowska, E.: Non-destructive testing of metal-polymer laminates with a digital holographic vibrometer. J. Nondestr. Eval. **39**(3), 1–8 (2020). https://doi.org/10.1007/s10921-020-00694-1

Performance of Selected Thermally Sprayed Coatings for Power Applications

Šárka Houdková, Zdeněk Česánek, and Pavel Polach(✉)

Research and Testing Institute Plzen, Tylova 1581/46, 30100 Plzeň, Czech Republic
`polach@vzuplzen.cz`

Abstract. In power industry, surfaces suffer from various kinds of degradation caused by external mechanical loading, increased temperature and corrosion aggressive environment. To protect the critical components' surface from such a degradation, various surface treatments are applied. The selected High Pressure/High Velocity Oxygen Fuel (HP/HVOF) sprayed coatings were tested in terms of mechanical, tribological and corrosion behavior and their performance was compared to the performance of the competitive types of surface treatment relevant to the aimed applications – a laser clad coating and a PVD thin film. The HP/HVOF sprayed coatings were chosen on the basis of their potential to resist wear, high temperature oxidation and corrosion in an aggressive environment. The completive types of surface treatment were chosen according to their relevance to a specific type of loading.

Keywords: Thermally sprayed coatings · HVOF · Wear resistance · High temperature · Corrosive environment

1 Introduction

The power generation industry is one of the most demanding areas in terms of material requirements. To protect the critical components surface from such a degradation, surface treatments are applied. With respect to the type of the component, its shape complexity and requirements given by the type of the loading, different types of surface treatment are chosen: surface hardening and nitriding for increased hardness in the case of low-temperature sliding wear suitable for small complex-shaped components, hard surfacing for the protection from oxidation, corrosion or wear, applicable on large components, where the temperature influence of a substrate material is not limitation, Physical Vapour Deposition (PVD) layers protecting the surface from wear, etc.

One of the applicable surface treatment technologies is a thermal spraying. This technology enables to create coatings from a wide range of materials including ceramics, metals and their alloys or hardmetals. Based on the used coating materials, different functionalities of the surface can be addressed. Thickness of the thermally sprayed coatings usually reached 0.2–0.3 mm, which is comparable to hard surfacing. The advantage of thermal spraying lies in the absence of the thermally affected zone in the underlying substrate. On the other hand, mechanical locking of the coating on the surface asperities does not provide the adhesive strength comparable to e.g. the laser or the electron beam cladding [1].

In the group of thermally sprayed technologies, High Velocity Oxygen Fuel (HVOF) offers the possibility to deposit coatings of metal alloys and hardmetals in a superior quality [1]. The kinetic energy of the coating materials´ particles impacting against the substrate is responsible for creating the coating of a low porosity and a high cohesive strength. Moreover, the relative low flame temperature (in comparison with e.g. atmospheric plasma spraying) keeps the undesirable phase changes in the coating material at a low level [2].

The hardmetal coatings are characterized by the combination of hard carbide particles with a tough metal matrix. While the carbides ensure the wear resistance, the matrix is mainly responsible for the corrosion and the oxidation resistance. Two types of hardmetal systems are widely spread: WC-based hardmetals suitable for low temperature applications up to 350 °C and Cr_3C_2-based hardmetal for the application up to 900 °C [3]. As the intended coatings application requires the oxidation resistance of the coating material in the hot steam environment (up to 610 °C), attention is paid to the Cr_3C_2-based hardmetals with 3 different types of matrix: NiCr; NiCrMoNb and CoNiCrAlY. From this group, the Cr_3C_2-25%NiCr composition is the most investigated one, both the HVOF and the Atmospheric Plasma Sprayed (APS) [4–7]. Up to now, less attention was paid to the coating of a variable matrix composition. The effect of thermal treatment on the tribological properties of Cr_3C_2-50%CoNiCrAlY was evaluated in [8]. Recently, the HVOF and the High Velocity Air Fuel (HVAF) sprayed Cr_3C_2-50%NiCrMoNb was analyzed in detail in [9].

For high temperature applications, the superalloys based on Co and Ni are used. A typical representative is a Co-based alloy known as Stellite®, the Hastelloy® Ni-based alloy or the NiCrBSi alloy. These materials often serve as coatings deposited using a laser cladding technology [10–12]. Alternatively, they can also be thermally sprayed, with the benefit of maintaining the original chemical composition without diluted zones typical for cladding.

The goal of the paper is to test the response of the above-mentioned coatings to the relevant loading conditions and to compare it to the alternative surface treatments to select the best surface treatments for specific applications. The coating materials were chosen on the basis of their potential to resist wear at a high temperature and in corrosive environment [13, 14]. The completive types of the surface treatment were chosen according to their relevance to the specific type of loading [15].

2 Materials and Testing Methods

The investigation involved the coating materials based on CrC-based hardmetals: Cr_3C_2-25%NiCr (Amperit 588.074); Cr_3C_2-25%CoNiCrAlY (Amperit 594.074); Cr_3C_2-50%NiCrMoNb (Amperit 595.074), Co-based alloys: CoCrW-Stellite (FST 484.33); CoCrAlTaCSi (Amperit 469.001) and Ni-based alloys: NiCrBSi (FST 341.33); NiCrMoW-Hastelloy C-276 (FST 771.33).

The coatings were deposited using the HP/HVOF TAFA JP5000 spraying device. The thicknesses of the sprayed coatings were set to 400 μm. For detailed information about the spraying process see [15–17]. The carbon steel substrates of dimensions in agreement with requirements of each test were grit blasted by Al_2O_3 (F22) media prior to the spraying to ensure coatings-substrate sufficient adhesion.

The types of competitive surface treatment were laser cladding of Co-based alloy (Stellite 6), PVD thin film TiAlN and gas nitrided X22CrMoV12-1 steel. Laser clad Stellite 6 coating was chosen for the comparison of microhardness HV 0.3, abrasive and sliding wear resistance and corrosion resistance in the aggressive 18% Na_2SO_4 82% $Fe_2(SO_4)_3$ environment. The PVD layer was tested and compared with the HP/HVOF sprayed coatings for hard particle erosion resistance and water droplet erosion resistance.

Evaluation of microstructures was performed on cross sections of coatings prepared by standard metallographic procedure. The microstructure was evaluated by optical (OM) and the scanning electron (SEM) microscopy. Using the HV 03 method, the microhardness was measured on the polished cross-sections of the coatings. At least 7 indents were made into each surface, from which the average value was calculated.

The Dry Sand/Rubber Wheel test was used to measure the abrasion resistance of three bodies in accordance with ASTM G65, using 22 N load, solid particle erosion was evaluated by the centrifugal erosion test [18]. Erosive media (Al_2O_3; F70) impacted at angles from 15° to 90°. For both tests, the mass loss was the starting point for determining the volume loss of the coatings using the density values previously obtained by the Archimedean method. Three independent measurements were made for each coating and the average value is given. SEM was used for subsequent observation of worn surfaces.

The Ball-on-Flat test was a criterion for evaluating the sliding wear resistance according to ASTM G133. The test parameters were as follows: AISI 440C steel; 25 N load; 5 Hz oscillating frequency; 6 mm diameter ball counterpart; 10 mm stroke length; 1,000 s testing time. For each coating were made three varied measurements. The KLA-Tencor P-6 Profiler profilometer was used as an instrument for measurement wear tracks' profiles at three different places and the rate of wear was calculated. The coating surface was ground and polished to the 0.04 ± 0.02 μm Ra value and subsequently subjected to sliding-wear tests.

The water droplet erosion test was performed in Doosan Škoda Power Ltd. Company in accordance with the company internal test prescription. The testing apparatus consists of a cylindrical vessel, in which a fixed strength disk is rotated, with the test pieces attached to the periphery. The nozzle creates a stream of droplets, through which the samples pass, causing the droplets to strike the samples at a prescribed velocity. Standard testing parameters are as follows: absolute pressure in the chamber: 3 kPa; shaft speed: 12,000 RPM; impact velocity: 524 m/s; droplet size: 0.41 mm; total test duration: 3 h 30 min. The samples were weighted in the defined period. The tested samples´ surface was smoothed prior to the testing using a metallographic procedure. Four samples of each type of coating were tested and the average value is reported.

Protective properties of the coatings in the high temperature aggressive environment of 18% Na_2SO_4 82% $Fe_2(SO_4)_3$ were tested. The thermal cycling from the room temperature up to 600 °C was repeated 50 times, the weight gain was recorded and evaluated. The high temperature test is described in detail in [14, 19].

2.1 Coatings Microstructure

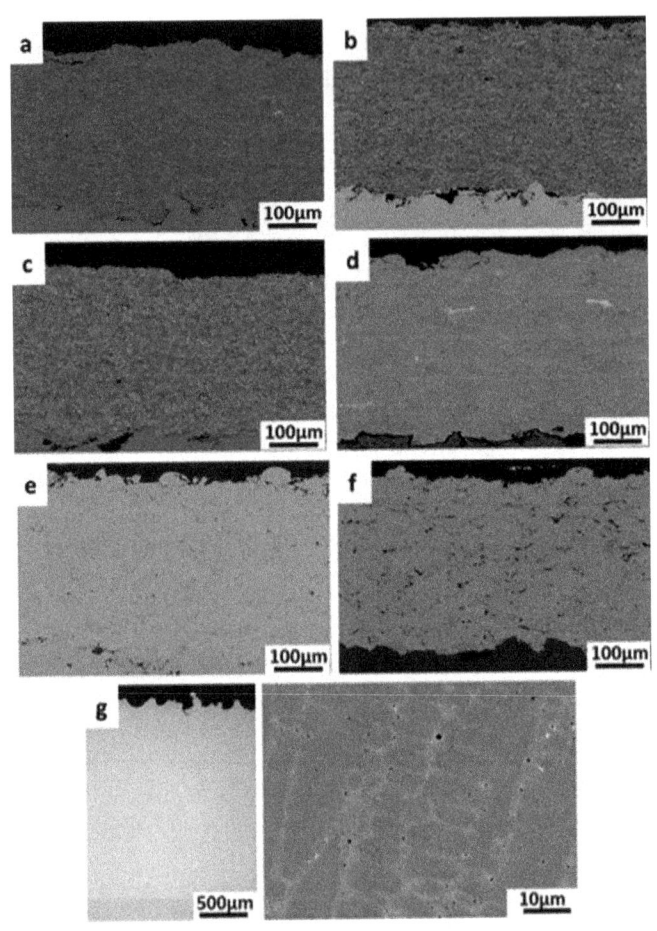

Fig. 1. Microstructure of coatings´ cross-sections: (a) Cr_3C_2-25%NiCr; (b) Cr_3C_2-25%CoNiCrAlY; (c) Cr_3C_2-50%NiCrMoNb; (d) Stellite HVOF sprayed coating; (e) NiCrBSi HVOF sprayed coating; (f) Hastelloy C-276 HVOF sprayed coating; (g) Stellite laser clad coating.

The coatings' cross-section microstructures can be seen in Fig. 1. No macro-defects, such as cracks or delamination, are visible in the observed coatings. The porosity is also negligible, except the Hastelloy C-276 coating (Fig. 1f), where pores are recognizable between the individual splats. The laser clad Stellite coating (Fig. 1g) has a dendritic structure consisting of the Co-based solid solution. Solitary small pores are concentrated in the interdendritic areas. As they do not connect each other, they do not represent a thread regarding the protection of the substrate from the corrosive environment.

2.2 Mechanical Properties Evaluation

Microhardness

The microhardness values measured on coatings cross-section are summarized in Fig. 2. While the Cr_3C_2-based hardmetal coatings and the NiCrBSi coating, whose microstructure is strengthened by fine hard boride and carbide particles, are comparable, the alloy HVOF sprayed Stellite and Hastelloy coatings are softer. The laser clad Stellite coating showed a lower hardness than the HVOF sprayed coating of a similar composition.

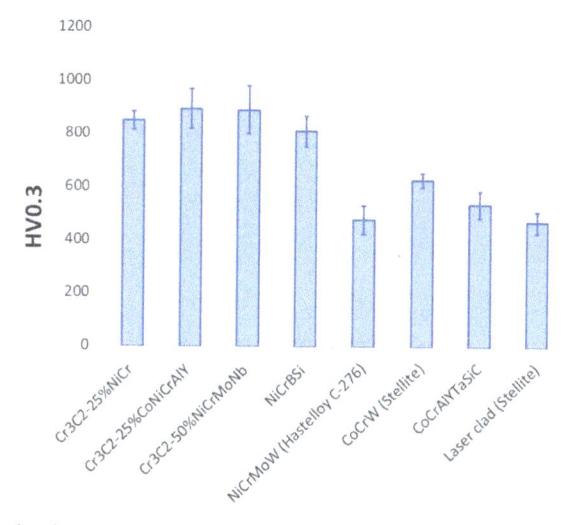

Fig. 2. Microhardness of the HVOF sprayed coatings and the laser clad stellite coating.

Abrasive Wear Resistance

Generally, the abrasion resistance of the HVOF coatings correlates with the microhardness values (Fig. 3). Hard CrC-based coatings show the highest abrasive resistance. From this group, the most resistant is the conventional Cr_3C_2-NiCr coating. Resistance of the Cr_3C_2-CoNiCrAlY coating is degraded by the lower cohesive strength of the coating, given by the high porosity and the weak bonding between the carbides and the surrounding matrix. In the case of the Cr_3C_2-50%NiCrMoNb coating, the lower resistance is a consequence of the higher matrix content. The NiCrBSi coating is of lower resistance, compared to the CrC-based coatings. Hardening particles in the coating (borides, carbides) are small in size compared to the abrasive particles and thus their effect is not

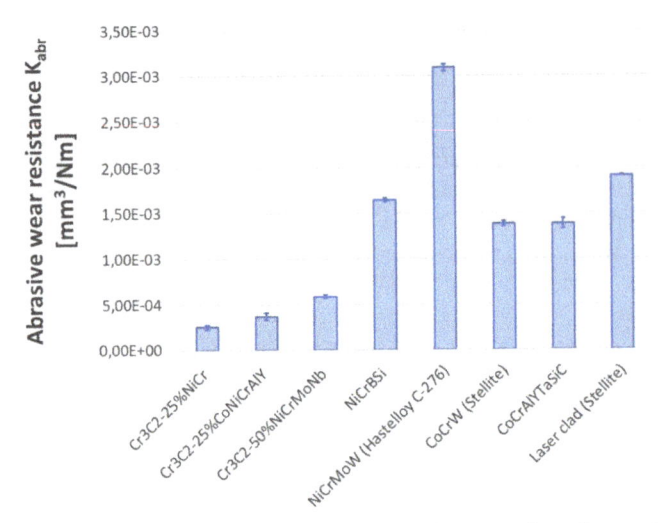

Fig. 3. Coefficient of abrasive wear resistance of the HVOF sprayed coatings and the laser clad Stellite coating.

as significant as the effect of larger carbides in the CrC-based coatings. The NiCrMoW coating alloy (Hastelloy C-276) consists only of a solid Ni solution without the presence of any reinforcing particles. Being the softest, its resistance to plastic deformation and wear is the lowest. Both Co-based coatings provide almost identical resistance to the abrasive wear. The laser clad Stellite coating was less abrasive resistant than the HVOF sprayed Stellite coating in agreement with its lower microhardness.

Sliding Wear Resistance

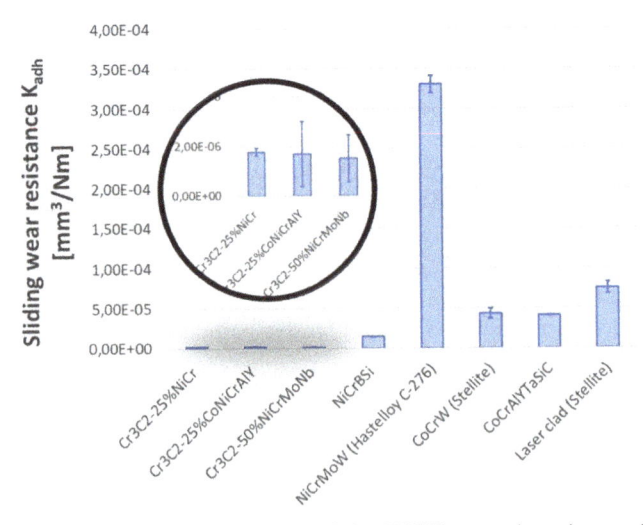

Fig. 4. Coefficient of sliding wear resistance of the HVOF sprayed coatings and the laser clad Stellite coating.

Sliding wear resistance of the carbide-based coatings is by an order of magnitude higher compared to the alloy coatings (Fig. 4). The CrC-based coatings sliding wear resistance is generally consistent with the data reported in other studies. The wear resistance of the most frequently evaluated Cr_3C_2-NiCr material ranges from $1 \cdot 10^{-6}$ to $1 \cdot 10^{-5}$ [4, 9], the value of $7.2 \cdot 10^{-6}$ was measured for the Cr_3C_2-CoNiCrAlY coating [8] and the value of $2.2 \cdot 10^{-6}$ [9] for the Cr_3C_2-NiCrMoNb coating, nevertheless these data originate in the pin-on-disc sliding wear test according to ASTM G99.

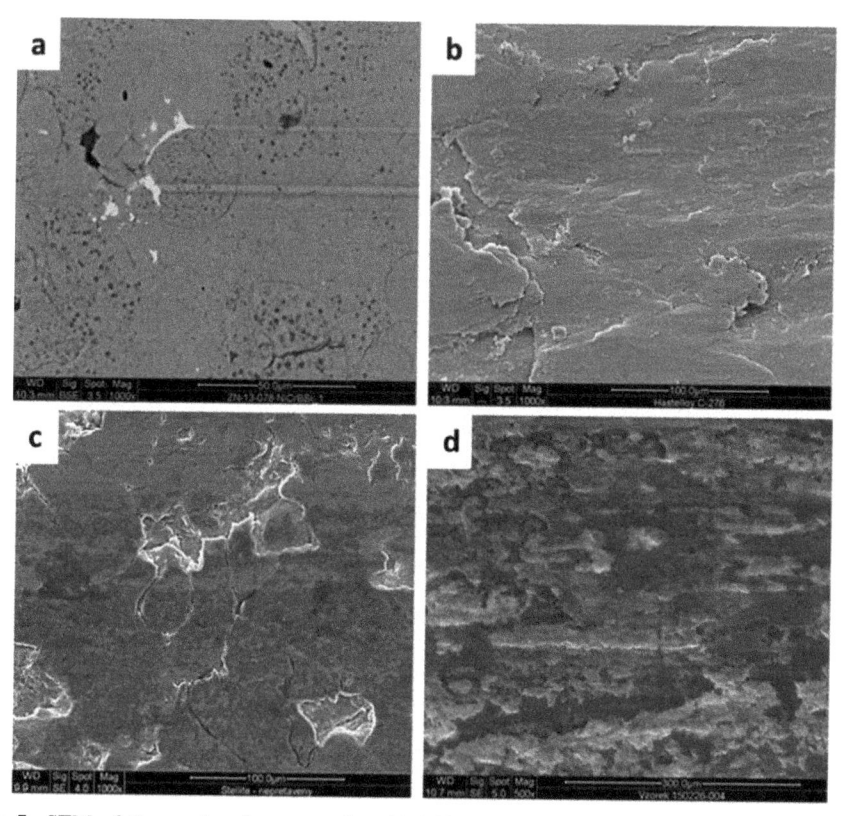

Fig. 5. SEM of the coatings´ wear tracks: (a) NiCrBSi HVOF sprayed coating; (b) Hastelloy C-276 HVOF sprayed coating; (c) Stellite HVOF sprayed coating; (d) Stellite laser clad coating.

Significant difference can be seen between two Ni-based alloy coatings. While NiCrBSi benefits significantly from the presence of fine reinforcing particles, the soft NiCrMoW coating is subject to the massive wear loss. The performance of the two Co-based coatings is similar as in the case of abrasive wear. The performance of the laser clad Stellite is also weaker compared to the HVOF sprayed Stellite coating. During the interpretation of the Co-based coatings wear resistance results the issue of deformation-induced transformation of which must be taken into account. It was discussed in the previous work [15] in more detail.

The difference in the sliding wear resistance is closely related with the differences in wear mechanisms. The wear mechanism in the CrC-based coatings was described many times. The cracks of carbides, their release from the matrix and the consequent cutting of matrix are the most frequent. In some cases, delamination of a larger part of the coating and the associated adhesive wear mechanism were observed [4].

The wear mechanism variability is demonstrated in Fig. 5. While only a groove from the released hard particle can be seen in the wear track of NiCrBSi coating (Fig. 5a), a massive plastic deformation and ploughing of NiCrMoW is responsible for high wear rate. While a delamination of whole splats appeared in the case of the HVOF Stellite coatings (Fig. 5c), a significant adhesive wear can be observed in the laser clad ones (Fig. 5d).

Erosion Wear Resistance

The solid particle erosion test reveals the poorer erosion resistance of the brittle CrC-based coatings while ductile alloy coatings showed from a lower erosion (Fig. 6). The CoCrW coating and the NiCrMoW coating achieved the best results for all angles of the erodent impact. In the case of a perpendicular impact of the erodent, the dominant properties of the coating are particularly the toughness and the cohesive strength of the splats, while the impact at an angle rather simulates the abrasive wear. Therefore, hard and brittle coatings break when the erodent strikes perpendicularly, while tough coatings are able to accumulate the impact energy in a better way. Although the erosion tests of the laser clad Stellite coating were not provided a high erosion resistance could be expected based on the previous experience.

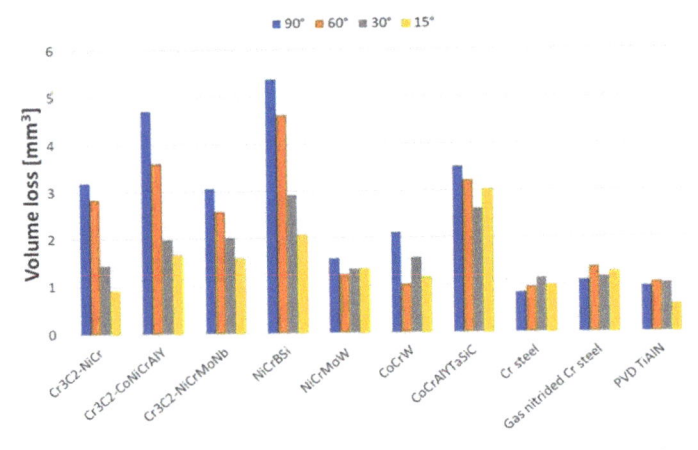

Fig. 6. Solid particle erosion of the selected HVOF sprayed coatings in comparison to the bulk Cr chrome and the PVD TiAlN layer.

Comparing to the base material and the PVD TiAlN layer, the HVOF sprayed coatings are less erosion resistant. It is caused probably by a lower cohesive strength of the coating, where delamination of individual splats can appear as a result of repeated impacting of hard erosive particles. Similarly, the water droplet erosion revealed poorer resistance of the HVOF sprayed coating compared to the basic material or the PVD layer under such a type of loading (Fig. 7, Fig. 8).

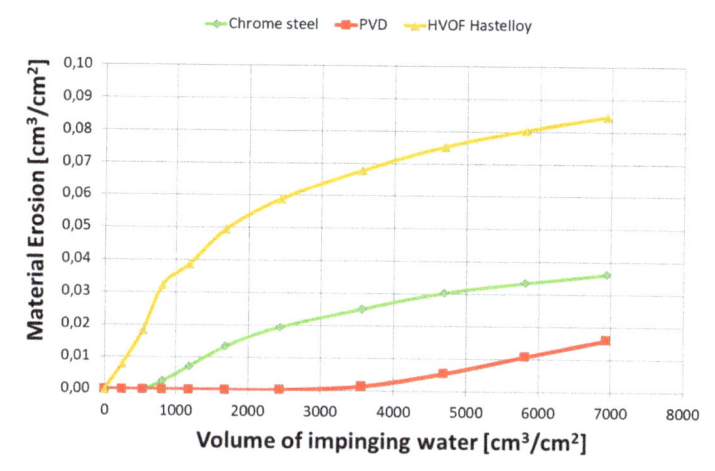

Fig. 7. Water droplet erosion of the selected HVOF sprayed coatings in comparison to the bulk Cr chrome and the PVD TiAlN layer.

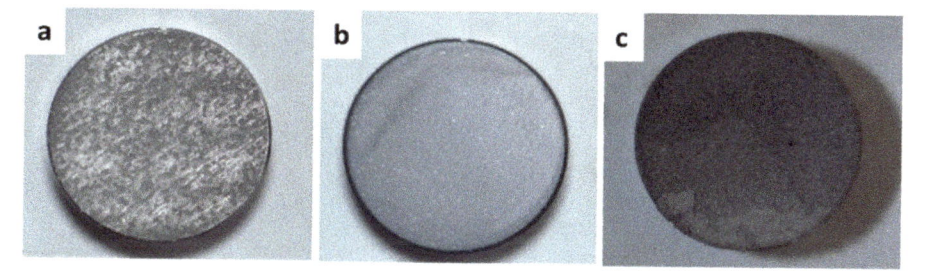

Fig. 8. The surface of eroded surfaces after 130 min of the water droplet erosion testing: (a) base material; (b) PVD TiAlN layer; (c) HVOF sprayed Hastelloy C-276 coating.

High Temperature Corrosion Test in the Aggressive Environment

Results of the high temperature corrosion testing in the aggressive environment of 18% Na_2SO_4 82% $Fe_2(SO_4)_3$ are shown in Fig. 9. This environment simulates real conditions in the combustion boilers of thermal power plants. All the tested HVOF sprayed coatings proved their ability to protect the underlying substrate and improve their lifetime. Their protective ability was even better than that of the laser clad Stellite 6 coating (Fig. 10).

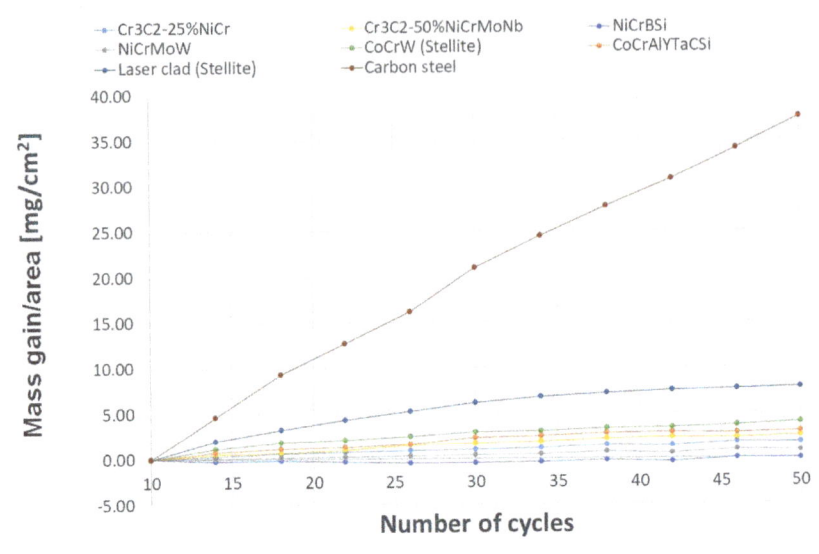

Fig. 9. Cumulative mass gain in dependence on the number of cycles at the high temperature corrosive aggressive environment.

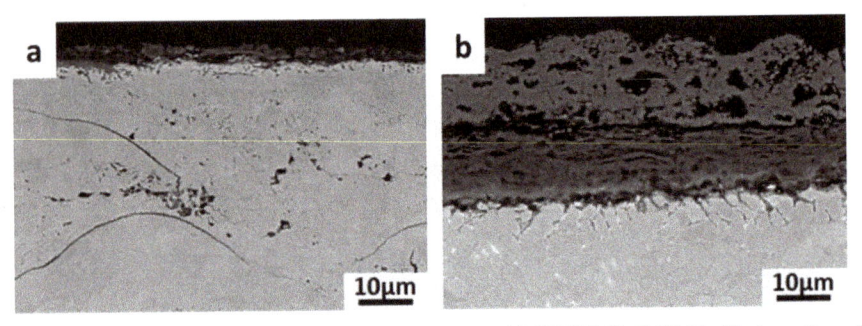

Fig. 10. SEM of the corroded coatings cross-sections: (a) HVOF CoCrW (Stellite) coating; (b) laser clad CoCrW (Stellite) coating.

3 Conclusions

Based on the above reported results, the superior behavior of the HVOF sprayed coatings, namely the Cr_3C_2-based hardmetals, was proved under the abrasive and the sliding wear loadings. On the other hand, the lamellar structure of the thermally sprayed coatings consisting of individual splats weakens the resistance of the HVOF coating under repeated loading, such as the solid hard particle erosion or the water droplet erosion. In this type of loading, the alloy coatings provide better results than the hardmetal coatings, but lower than the alternative PVD layer. Regarding the protection of the substrate from the corrosive environment they are comparable to the laser clad Stellite coatings. Based on these findings, the application on the components suffering from oxidation and corrosion can be recommended if the parts are mostly loaded with sliding or abrasive wear.

Acknowledgement. The paper has originated in the framework of institutional support for the long-time conception development of the research institution provided by the Ministry of Industry and Trade of the Czech Republic to Research and Testing Institute Plzen.

References

1. Pawlowski, L.: The Science and Engineering of Thermal Spray Coatings, 2nd edn. Wiley, New York (2008)
2. Vaidya, A., Streibl, T., Li, L., Kovarik, O., Greenlaw, R.: An integrated study of thermal spray process structure – property correlations: a case study for plasma sprayed molybdenum coatings. Mater. Sci. Eng. A **403**(1–2), 191–204 (2005)
3. Berger, L.M.: Application of hardmetals as thermal spray coatings. Int. J. Refract. Metal. Hard Mater. **49**, 350–364 (2015)
4. Bolelli, G., et al.: Sliding and abrasive wear behaviour of HVOF- and HVAF-sprayed Cr_3C_2-NiCr hardmetal coatings. Wear **358–359**, 32–50 (2016)
5. Espallargas, N., Berget, J., Guilemany, J.M., Benedetti, A.V., Suegama, P.H.: Cr3C2-NiCr and WC-Ni thermal spray coatings as alternatives to hard chromium for erosion-corrosion resistance. Surf. Coat. Technol. **202**(8), 1405–1417 (2008)
6. Hussainova, I., Pirso, J., Antonov, M., Juhani, K., Letunovitš, S.: Erosion and abrasion of chromium carbide based cermets produced by different methods. Wear **263**, 905–911 (2007)
7. Matthews, S.: Development of high carbide dissolution/low carbon loss Cr_3C_2-NiCr coatings by shrouded plasma spraying. Surf. Coat. Technol. **258**, 886–900 (2014)
8. Picas, J.A., Punset, M., Menargues, S., Martín, E., Baile, M.T.: Microstructural and tribological studies of as-sprayed and heat-treated HVOF Cr3C2-CoNiCrAlY coatings with a CoNiCrAlY bond coat. Surf. Coat. Technol. **268**, 317–324 (2015)
9. Matikainen, V., Bolelli, G., Koivuluoto, H., Lusvarghi, L., Vuoristo, P.: Sliding wear behaviour of thermally sprayed Cr3C2-based coatings. Wear **388–389**, 57–71 (2017)
10. Frenk, A., Vandyoussefi, M., Wagnière, J.-D., Kurz, W., Zryd, A.: Analysis of the laser-cladding process for stellite on steel. Metall. Mater. Trans. B. **28**, 501–508 (1997)
11. Serres, N., Hlawka, F., Costil, S., Langlade, C., Machi, F., Cornet, A.: Dry coatings and ecodesign part. 1 — environmental performances and chemical properties. Surf. Coat. Technol. **204**(1–2), 187–196 (2009)
12. d'Oliveira, A.S.C.M., da Silva, P.S.C.P., Vilar. R.: Microstructural features of consecutive layers of stellite 6 deposited by laser cladding. Surf. Coat. Technol. **153**, 203–209 (2002)
13. Sidhu, T.S., Prakash, S., Agrawal, R.D.: Characterization and hot corrosion resistance of Cr_3C_2-NiCr coating on Ni-base superalloys in an aggressive environment. J. Therm. Spray Technol. **15**, 811–816 (2006)
14. Česánek, Z., Houdková, Š, Lukáč, F.: High-temperature corrosion behavior of selected thermally sprayed coatings in corrosive aggressive environment. Mater. Res. Exp. **6**(1), 016426 (2019)
15. Houdková, Š, Pala, Z., Smazalová, E., Vostřák, M., Česánek, Z.: Microstructure and sliding wear properties of HVOF sprayed, laser remelted and laser clad stellite 6 coatings. Surf. Coat. Technol. **318**, 129–141 (2017)
16. Houdková, Š, Česánek, Z., Smazalová, E., Lukáč, F.: The high-temperature wear and oxidation behavior of CrC-based coatings. J. Therm. Spray Technol. **27**, 179–195 (2018)

17. Houdková, Š, Smazalová, E., Vostřák, M., Schubert, J.: Properties of NiCrBSi coating, as sprayed and remelted by different technologies. Surf. Coat. Technol. **253**, 14–26 (2014)
18. Deng, T.: A comparison of the gas-blast and centrifugal-accelerator erosion testers: the influence of particle dynamics. Wear **265**, 945–955 (2008)
19. Česánek, Z., Houdková, Š., Lencová, K., Schubert, J., Mušálek, R., Lukáč, F.: High temperature corrosion behavior of selected thermally sprayed alloy based coatings in aggressive environment at 690 °C. In: Krömmer, W., McDonald, A., Bobzin, K., Lenling, W., Ogawa, K. (eds.) International Thermal Spray Conference and Exposition ITSC 2019: New Waves of Thermal Spray Technology For Sustainable Growth on Proceedings, pp. 214–221. ASM International, Yokohama, Japan (2019)

Laser Measurement by Angle Accuracy Method in Additive Technology SLM 316L

Ondrej Mizera[✉], Jiri Hajnys, Lenka Cepova, Jan Zelinka, and Jakub Mesicek

Department of Machining, Assembly and Engineering Metrology,
Faculty of Mechanical Engineering, Technical University of Ostrava, 17. listopadu 2172/15,
708 00 Ostrava-Poruba, Czech Republic
{ondrej.mizera,jiri.hajnys,lenka.cepova,jan.zelinka,
jakub.mesicek}@vsb.cz

Abstract. This document examines the manufacturing accuracy of AISI 316L corrosion steel material with 3D printing across geometric tolerances with a focus on angle formation with and without support. The measured data were compared with a CAD model, which also served as an input element to a 3D printer. The experiment points out the possibilities of creating angles without support up to certain degrees and their deviation from the ideal body.

Keywords: Additive manufacturing · Steel AISI 316L · Laser method measurment

1 Introduction

3D printing technology is a modern science that works on the principle of melting materiál in the from of powder and applying layer by layer in the height axis Z, to create a finished product. Using the selective laser melting (SLM) methods, geometrically complex components with higher mechanical properties are produced compared to ypical production by machining [1, 2]. Additive technology includes, for example, sintering or sintering of powders of various chemical composition, grain size, physical, chemical and other useful properties. Directly due to is character advantages, the SLM method is the most widely used method of metal powder printing, where it has found i tis presence in many scientific fields such as the aerospace industry, medicine, automotive or energy, and many others [3, 4]. A typical unfavorable phenomenon for the SLM method is the high temperature of the transition areas between each layer. This phenomenon manifests itself in the final product as poor surface microstructure or unmet requirements and mechanical properties, this has a direct effect on the surface integrity for 3D printing [5–7]. When it comes to design parts and not for functional use, the most common use is the tumbling process, for parts with functional use and exact dimensional values and geometric tolerances, the parts are machined. These post-process operations serve to improve surface size and roughness. Only the tumbling method is used for the complex surface treatment of printed components, there is a lot of experimental study on this topic and the surface roughness of printed components is a very rich topic in the category of additive technology. Tumbling itself is very time consuming and economically raises the cost of components [8].

In order to ensure the functionality and required accuracy of products made with additive technology, it is necessary to specify the methodology, types of meters and measuring devices that are able to cover the requirements determining the range of required tolerances, or selecting the most optimal measurement methods for the purpose. In our case, the optical method was chosen, a type of laser scanner that can record 48,000 data points per second with an accuracy of 20 µm [9] (Fig. 1).

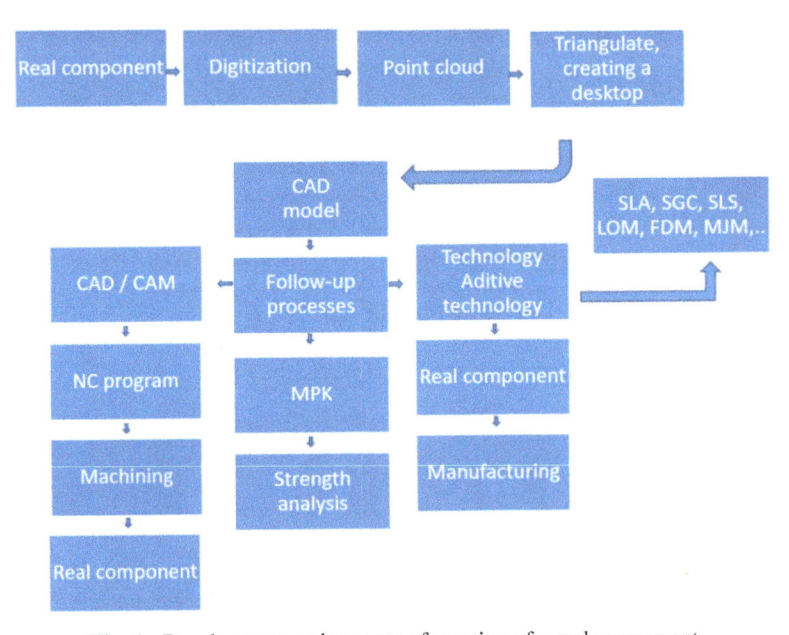

Fig. 1. Development and process of creation of a real component.

In order to ensure the functionality and required accuracy of products made with additive technology, it is necessary to specify the methodology, types of meters and measuring devices that are able to cover the requirements determining the range of required tolerances, or selecting the most optimal measurement methods for the purpose. In our case, the optical method was chosen, a type of laser scanner that can record 48.000 data points per second with an accuracy of 20 µm [9].

The experiment was focused on the material AlSi 316L with the detection of production accuracy in comparison with the CAD model, where the geometric deviations will be evaluated, specializing in angles and their creation in the working chamber of a 3D printer [10–12]. Investigation of the influence of angle formation in various spaces and print positions. Additive metal powder technology allows us to create external and internal shapes of components of any complexity, which ultimately brings direct production of complex and geometrically complex parts - making parts at once and in many pieces in one work chamber, saving production costs, shortening assembly time, increasing reliability, etc. [12]. The work experiment examined only the accuracy of production in comparison with the CAD model. Mrs. Monková and Mr. Monka [13] from TUKE Košice dealt with qualitative parameters of a complex component made by an additive approach, where from the beginning of the study it was clear that the surface will not be smooth with low surface roughness due to the principle of technology. Consistent with this prediction, the results also showed relatively large differences between the parameters measured at different points in the study areas. In order for 3D printing technology to be part of the industrial production of real components, it is necessary to achieve high quality properties of manufactured parts. Based on the above, it is clear that the final properties (quality and accuracy) of the product strongly depend on the production speed, part orientation, layer thickness and need for support structures, cut thickness, assembly orientation, support structures and hatching pattern [13]. For example, Allen and Dutta [14] see Chapter 2, deal with the issue of component placement in the working chamber and their exact location in the working chamber, the placement of components has a great influence on its geometric and mechanical properties. It is an effort to achieve as few supports on the component as possible so that it does not bend due to internal stress during removal of supports.

This experiment was focused on the geometric accuracy of production with a focus on radius production. We investigated to which angular sizes we do not need to create internal supports and we can use internal supports in the form of powder in the chamber of the surrounding substrate and from which angular sizes we need to create external fixed supports, see Fig. 3, which must be cut after the production process.

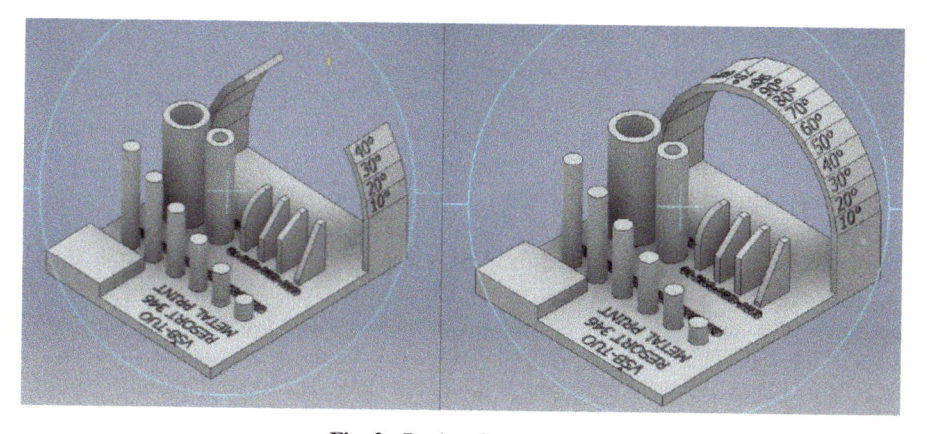

Fig. 2. Design CAD model.

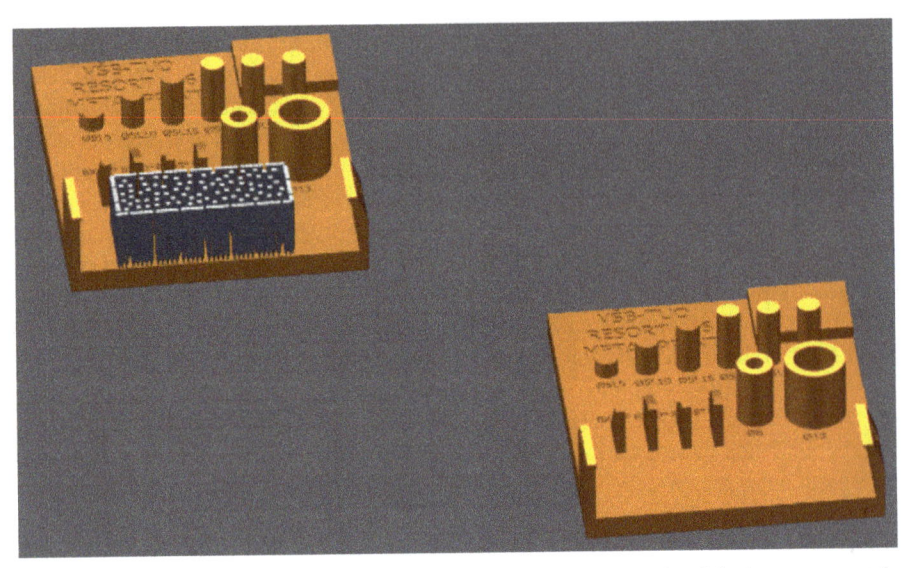

Fig. 3. From the left is support for the production of arch and from the right is not support for product.

2 Fitting Best Fit

When inspecting the manufactured part, it cannot be assumed that we will scan it and automatically obtain the result from the program with deviations for comparison with the CAD model. Inspection programs cannot know exactly how a given part is to be manufactured, what its functional properties are and which points are important for the correct placement of a part in the overall assembly. For these reasons, it is essential that the inspector correctly selects points at the beginning of the part inspection to create a blend of the scanned data relative to the CAD model. Subsequent deviations will reflect reality. The most commonly used alignment is BEST Fit, which seeks the best fit of the shape of the scanned and CAD model on selected (local) surfaces corresponding to each other on the scanned and reference model. In this case, it is necessary to use the machined surfaces of the resulting part; for surface casting without machining, significantly higher deviations can be expected [15]. This alignment is used in cases where we do not know what the starting points are in the work. The program seeks to achieve minimum deviations at all points on both models by aligning the measured points or a set of real elements matching as close as possible to its nominal position or theoretical counterpart. In some cases, it is also possible for the Best Fit alignment to optimally match the clouds of the scanned points to the CAD curve or surface [16].

After fitting to the selected points, there will be an overlap of both models with color transitions. The richer the color, the closer it is to the deviation value. The color scale of deviations is fully adjustable in terms of the size of the deviation (mm), as well as the color design of the thermographic map, which is colored different colors from the size of the deviation [16, 17]. This graphic determines the deviation from the actual model.

In addition to the color spectrum, another type of display often uses a certain type of isoline [17, 18].

The Best Fit least squares algorithm aligns two sets of points by transforming one of the sets so that the sum of the squares of the distances between the corresponding points in the two sets is minimal [17, 18].

All experimental work will take place on a 3D CMM Wenzel LH 65 X3Premium with a laser head SHAPETRACER II from Wenzel (Fig. 4).

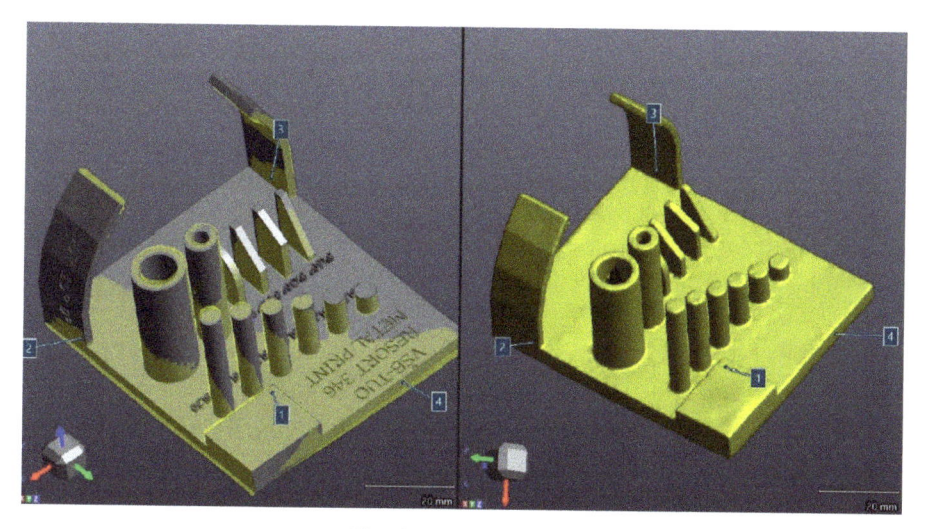

Fig. 4. Best fit method.

3 Part Orientation

Messrs. Masood and Rattanawong [19] have developed an algorithm for calculating the magnitude of the volume error caused by cutting a CAD model. Volume errors are calculated for various rotations around the specified axes. The best orientation is then determined as the component with minimal volume error [19, 20]. At the University of West Bohemia in Pilsen, Mrs. V. Čapková [20, 21] conducted research focusing on the limiting parameters of 3D printing technology used for the production of metal components. The aim of this work is to collect as many critical parameters as possible and perform their theoretical analysis. Depending on these limiting parameters, a component was designed that contains a number of geometric elements on which the critical parameters were measured [22]. The position of the component or its orientation during construction has a great influence on its accuracy, surface quality, construction time, the required number of support structures and also on the costs associated with production. Therefore, the most appropriate orientation must be determined with respect to the various parameters in order to achieve all the requirements that the user prefers. Probably the most important criterion when choosing an orientation is the final quality of the printed part. As already mentioned, this is an additive method of production, so there is often a so-called stair effect, which affects the quality of the final assembly. This effect most often occurs on curved and inclined surfaces and adversely affects their dimensional accuracy [20–22]. For example, Messrs. Allen and Dutta developed a method for the automatic calculation of the load-bearing structure and a decision method for the most appropriate choice of orientation, which was chosen from a selected list of different orientations. Frank and Fadel designed an expert system that takes into account various parameters that affect the production of a component. This system works with the specified input parameters and the created decision matrix, according to which the system will recommend the most suitable direction of construction and orientation of the component in the workspace.

The part was oriented during production (see Fig. 5 - orientation of production in the construction room) and both versions of the part with support and without support (see Fig. 2 - Design CAD model) orientation was different to find out what effect the orientation of the formed element for geometric deviations of shape and position. During the experiment, we focused only on the evaluation of the arc and the evaluation of the angle. This arc element had the largest geometric deviations in production compared to the CAD model. The CAD model served as a standard for evaluation.

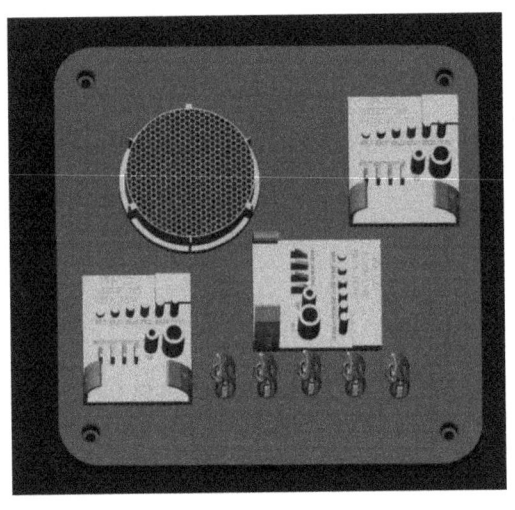

Fig. 5. Orientation of production in the construction chambre.

4 Experiment Production

The construction was carried out chronologically according to the technological production sequence of additive production according to the standard ČSN EN ISO 17 296-4 [23]. Volume models in the computer program Inventor Professional were processed for the production of components [24]. Due to the experimental design of the component, the technological maximum of printing by the SLM method was reached. During the printing of the arc, destructive defects began to occur, at a plane angle in the range from 40° to 50° of curvature. It was necessary to introduce static support into the vault of the arch. The supports ensured stabilization of the print for the full arc model (see Fig. 3 - From the left is support for the production of arch and from the right is not support for product). Technological parameters for printing have been changed for printing stability. The initial plan was to set the beam power to 400 W when printing the first part. The laser power was adjusted to a final value of 200 W due to insufficient printing. The printing was stabilized for other new parts after correction of the laser power. The printing speed of individual sections of experimental components was 650 mm/s. The construction height of the higher part is 49.98 mm, the model is printed in individual layers after 50 μm, the cutting part was divided into 999 layers of printing. The total printed volume of both components with supports is 88077 mm^3 (Figs. 6 and 7).

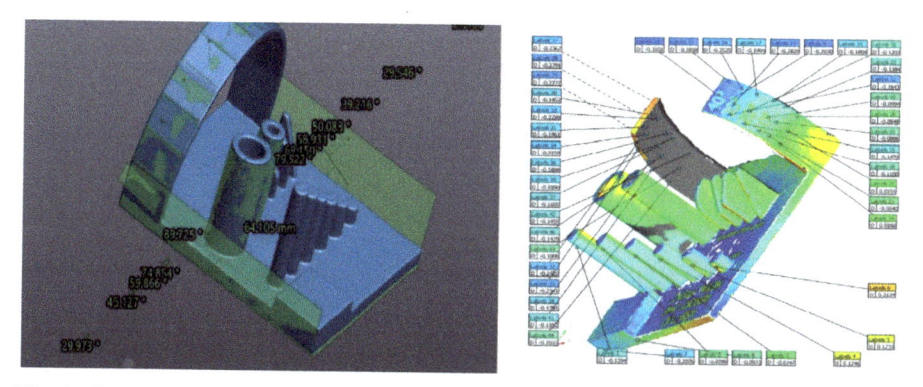

Fig. 6. Geometric deviations with the marking of the thermographic map of the part with full arc.

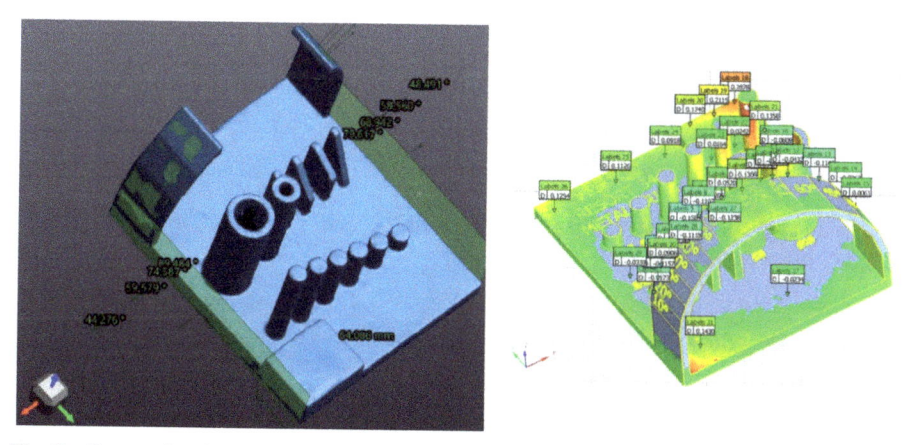

Fig. 7. Geometric deviations with the marking of the thermographic map of the half-circumference part.

At the interface of 30° to 40°, the tolerance for an angular deviation of 0.421° is positive 0.176 mm (Tables 1, 2 and 3).

Table 1. Angle values on partially arced parts.

Angle	Angle on the part	Deviation	Measured angle	Angle	Angle on the part	Deviation	Measured angle
10°	10.383°	0.383°	79.617°	0°	0.516°	0.516°	89.484°
20°	21.058°	1.058°	68.942°	15°	15.413°	0.413°	74.587°
30°	31.440°	1.440°	58.560°	30°	30.421°	0.421°	59.579°
40°	41.509°	1.509°	48.491°	45°	45.724°	0.724°	44.276°

Table 2. Deviations of measured values on components with a partial arc of the device.

Measurement number (labels)	Deviation	Measurement number (labels)	Deviation
19	0.0994 mm	25	0.0155 mm
16	0.1203 mm	43	−0.1006 mm
13	0.2829 mm	40	−0.1479 mm
10	0.3101 mm	28	−0.2797 mm

Table 3. Angle values on parts with a full arc.

Angle	Angle on the part	Deviation	Measured angle	Angle	Angle on the part	Deviation	Measured angle
10°	10.478°	0.478°	79.522°	0°	0.275°	0.275°	89.725°
20°	20.850°	0.850°	69.150°	15°	15.146°	0.146°	74.854°
30°	31.069°	1.069°	58.931°	30°	30.134°	0.134°	59.866°
40°	39.917°	0.083°	50.083°	45°	44.873°	0.127°	45.127°
60°	60.454°	0.540°	29.546°	60°	60.027°	0.027°	29.973°

For an angle of 60°, the angular deviation is 0.027° and is also positive with a measure of 0.357 mm (Table 4).

Table 4. Deviations of measured values on components with a complete arc of the device.

Measurement number (labels)	Deviation	Measurement number (labels)	Deviation
27	−0.1238 mm	19	−0.2115 mm
21	0.1358 mm	26	−0.1254 mm
18	0.2878 mm	13	−0.1146 mm
32	0.0935 mm	8	−0.1107 mm

5 Conclusion

Great influence on the quality of the process of adjusting the power of the laser beam from 400 W to 200 W. By fitting the part with the BEST FIT method, the body periods (attempting to place the same points) for the circulating components were selected. Areas of tolerance differences are symmetric for circulating methods. There were 3 pieces of each type and they were compared. A model with a partial arch of the arch is more suitable for expressing geometric tolerances. The measuring device was detected for an area with a 45° angle of negative deviation from the CAD model – 0.2797 mm. The static support used for printing solid arc parts is highly reflected in the degree of deviation from the CAD model. A part with a full arc is also a precise production made than a part with a partial arc. The deviation values on surfaces at an angle of 30° are 1.440° and 0.421° for parts with a partial arc. For a part with a full arc on surfaces with an angle of 30°, the measured deviations take the values of 1.069° and 0.134°.

5.1 Plan

We plan with the team to continue research into the creation of the angle of additive technology using the SLM method, the research will be integrated into the metrological and mechanical fields.

Funding. This paper was completed in association with the project Innovative and additive manufacturing technology—new technological solutions for 3D printing of metals and composite materials, reg. no. CZ.02.1.01/0.0/0.0/17_049/0008407 financed by Structural Funds of the European Union and project.

This paper article has been elaborated in the framework of the grant programme "Support for Science and Research in the Moravia-Silesia Region 2018" (RRC/10/2018), financed from the budget of the Moravian-Silesian Region.

References

1. Frazier, W.E.: Metal additive manufacturing: a review. J. Mater. Eng. Perform **23**, 1917–1928 (2014)
2. Hlinka, J., et al.: Complex corrosion properties of aisi 316L steel prepared by 3D printing technology for possible implant applications. Materials **13**(7) (2020). https://doi.org/10.3390/ma13071527
3. Gao, W., et al.: The status, challenges, and future of additive manufacturing in engineering. Comput. Aided Des. **69**, 65–89 (2015)
4. Haleem, A., Javaid, M.: 3D printed medical parts with different materials using additive manufacturing. Clin. Epidemiol. Glob. Health (2019)
5. Kong, D., et al.: Heat treatment effect on the microstructure and corrosion behavior of 316L stainless steel fabricated by selective laser melting for proton exchange membrane fuel cells. Electrochim. Acta **276**, 293–303 (2018)
6. Lichovník, J., Mizera, O., Sadílek, M., Čepová, L., Zelinka, J., Čep, R.: Influence of tumbling bodies on surface roughness and geometric deviations by additive SLS technology. Manuf. Technol. **20**(3), 342–346 (2020). https://doi.org/10.21062/mft.2020.050

7. Kořínek, M., et al.: Monotonic tension-torsion experiments and FE modeling on notched specimens produced by SLM technology from ss316L. Materials **14**(1), 1–15 (2021). https://doi.org/10.3390/ma14010033
8. Mesicek, J., et al.: Abrasive surface finishing on SLM 316L parts fabricated with recycled powder. Appl. Sci. (Switzerland) **11**(6) (2021). https://doi.org/10.3390/app11062869
9. Strnadel, F.: Methods of 3D CMM measurement of selected components of additive manufacturing. Bachelor thesis, 65 (2021)
10. ČSN EN ISO 1101 (014120): Geometric Product Specifications (GPS): Geometrically tolerated: Tolerances on shape, orientation, location and throw. Institute for Technical Standardization, Metrology and Testing, Prague (2017)
11. ČSN ISO 2768–2 (014406) General tolerances: Non-prescribed geometric tolerances. Institute for Technical Standardization, Metrology and Testing, 12 (1994)
12. Additivní technology - Method Rapid Prototiping. http://ust.fme.vutbr.cz/obrabeni/poklady/sto_bak/cv_STV_04_Aditivni_technologie_metody_Rapid_Prototyping.pdf. Brno (cit. 17 June 2021)
13. Monkova, K., Monka, P.: Qualitative parameters of complex part produced by additive approach. Paper presented at the 2017 8th International Conference on Mechanical and Aerospace Engineering, ICMAE 2017, pp. 691–694 (2017). https://doi.org/10.1109/ICMAE.2017.8038732
14. Allen, S., Dutta, D.: About the calculation of the orientation of the part using the supporting structure in layered production. In: Solid Freeform Fabrication Symposium, pp. 59–269 (1994)
15. Component shape inspection. Ostrava [cit. 1 Sept 2021]. http://www.iae.fme.vutbr.cz/userfiles/ramik/files/Studium/Poklady%20ke%20studiu/3D%20skenov%C3%A1n%C3%AD/E-learning-Inspekce%20tvaru%20soucasti.pdf
16. Alignment used for quality control (QC). Brno: 3dSCAN (2020) [cit. 14 June 2021]. https://www.3d-skenovani.cz/pouzivane-zarovnani-pri-kontrole-kvality-qc/
17. https://www.wenzel-group.com/en/products/wenzel-shapetracer-laser-scanner/
18. https://support.hexagonmi.com/s/article/Understanding-Best-Fit-Alignments-1527312042816
19. Massod, S.H., Rattanawong, W., Lovenitti, P.: A generic algorithm for a best part orientation system for complex parts in rapid prototyping **139**(1–3), 110–116 (2003). https://doi.org/10.1016/S0924-0136(03)00190-0
20. Mishra, A.K., Thirumavalavan, S.: A study of part orientation in rapid prototyping (2014)
21. Čapková, V., Zetková, I.: RP technology used for the production of metal part. Mater. Sci. Forum **818**, 280–283 (2015)
22. Čapková, V.: Restrictive parameters of 3D printing (2015)
23. ČSN EN ISO 17 296-2 (011810): Additive manufacturing - Basic principles - Part 2: Overview of process and raw material categories (2017)
24. Čepova, L., Kovacikova, A., Cep, R., Klaput, P., Mizera, O.: Measurement system analyses – gauge repeatability and reproducibility methods. Measur. Sci. Rev. **18**(1), 20–27 (2018). https://doi.org/10.1515/msr-2018-0004

Review of Measurement Methods to Evaluate the Geometry of Different Types of External Threads

Bartłomiej Krawczyk[1,2(✉)], Krzysztof Smak[1,2], Piotr Szablewski[2,3], and Bartosz Gapiński[1]

[1] Faculty of Mechanical Engineering, Poznan University of Technology, 3 Piotrowo Street, 60-965 Poznan, Poland
bartlomiej.krawczyk@prattwhitney.com
[2] Pratt & Whitney Kalisz, 4a Elektryczna Street, 62-800 Kalisz, Poland
[3] Higher Vocational State School President Stanislaw Wojciechowski in Kalisz, 4 Nowy Świat Street, 62-800 Kalisz, Poland

Abstract. The article summarizes the literature review in terms of currently used measuring methods for the evaluation of selected features of external threads. The work focuses on high quality requirements for threads produced for the needs of the aviation industry. The authors noticed that the classic measurement methods are mainly manual, but in recent years attempts have been made to automate the control of this type of elements. Selected methods found in the world literature, both manual and automatic as well as contact and contactless, have been described and compared in this paper. The authors attempted to examine selected measurement methods on three test pieces with different thread profiles: triangular, trapezoidal symmetrical and trapezoidal asymmetrical. A coordinate measuring machine, an automatic optical measurement device and a three-wire method were used in the research. The obtained results are presented and briefly summarized. It can be confirmed that it is possible to automate the measurement of selected features of external threads. The need for further research has been noticed.

Keywords: External thread · Measurement · Quality control · Aerospace industry

1 Introduction

Threaded connections are the most commonly used detachable connections in the assembly process of mechanical parts and are practiced in the manufacturing and aerospace industries [1]. These extremely important constructional elements account for about 50% of all mechanical connections [2]. This connection is based on the cooperation of two threaded components and their mutual operation is possible due to the frictional forces occurring at the interface between the surface of the external thread and the internal thread. A thread is a ridge evenly spaced along a helix on the side surface of a cylinder or a straight rotating cone [3]. In addition to the assembly function, the threads can also be movable connections to transform a rotary movement into a linear movement and vice versa, or a force transmission to generate more force by using less.

The improvement of technological processes influenced the significant spread of the introduction of threaded connections in the construction of machines and devices. It should be noted that these mechanisms play an extremely important role not only in simple applications, but also are widely used in critical connections that undoubtedly occurs in the aviation industry, e.g. in jet engines. The specificity of these products requires the maintenance of the highest quality of all elements, as they are responsible for the correct operation of the aircraft propulsion. This allows to maintain the efficiency and reliability of the engine, and thus ensure the safety of passengers throughout the flight, when the parts are exposed to extreme changes in operating conditions. It should not be forgotten that apart from external conditions, there is also the aspect of fuel combustion inside the engine compartment where the temperature can reach almost 1600 °C [4]. For this reason, engine components are often made of difficult-to-machine, high-strength, heat-resistant alloys, e.g. nickel-based alloys (Inconel 718), cobalt, molybdenum [5, 6]. These materials, combined with high dimensional and shape requirements, makes the production of threads that meet safety demands definitely a difficult task, which forces the aviation industry to introduce and comply with extremely restrictive regulations and quality standards.

Therefore, companies from the aerospace industry use and constantly implement modern and innovative solutions for machines, tools and machining methods. In recent years, a growing problem of the manufacturing industry is the lack of qualified employees, so this is the reason why continuous improvement of processes, and automation are becoming indispensable activities which allow to remain competitive on the global market. In modern production plants, one of the key goals in recent years is the implementation of the Industry 4.0 concept, according to which machining processes are performed in the technology called "closed door machining", i.e. without operator interference [7, 8]. This action allows to increase the production efficiency, and also ensures greater repeatability, which in turn affects the quality of the final products. Threaded elements are one of such products. However, in addition to the commonly known metric and inch connections, aero-engines contain more sophisticated symmetrical and asymmetrical special solutions, e.g., buttress threads. The quality and precision requirements for this type of product are much higher than in the case of traditional threads used when joining machine parts. For example, the tolerance of the pitch diameter of an external thread on an engine shaft made of Inconel 718 could be ± 50 μm. This results in the need for precise control of such products, and following the philosophy of "closed door machining", the control process should also take place without operator intervention. This would allow to shorten the measurement time, increase its repeatability and eliminate the fallibly human factor. Due to the specificity of this products, removing them from a numerically controlled lathe practically excludes, in production conditions, the introduction of any corrections.

For this reason, the measuring process is increasingly becoming an integral part of the manufacturing process. If it is possible to implement it directly on the machine, process time is reduced by eliminating transport to the measuring station, costs are reducing by eliminating measuring stations and the possibility of direct product corrections appear.

Moreover, if the measurement process is carried out in automatic mode, then the negative influence of the human factor on the correctness of the obtained results is eliminated. Solutions of this type are found in relation to simple geometric characteristics (and this in a much smaller range than measurements carried out on a coordinate measuring machine in laboratories), however, in the case of making threads on a CNC machine, measurements of this type have not been carried out so far.

This paper is an introduction to further considerations on the selection of an appropriate measuring method and integration it with the machining process, which would allow measuring turned threads in automatic way directly on a CNC lathe.

2 Methodology of External Threads Measurement

Due to the fact that the thread is described by a number of dimensions, considering the measurement methods, it should first of all be specified which of the geometrical features requires control. Thread measurement methods can be classified in two ways. Depending on the interpretation, there are manual and automated measurements as well as contact and non-contact measurements. Typical methods described in classic handbooks are based on devices that require operator support. When analyzing scientific articles from recent years, it can be noticed that attempts are made to automate thread measurements. In many cases, described by researchers, measurement methods have been developed that allow to obtain a result with an acceptable measurement uncertainty.

It is known, that the range of dimensional tolerances of threads has been defined in general standards. Sometimes, as already mentioned in the introduction, these requirements may be tightened due to the use of the product in a responsible application. In the case of critical connections, high emphasis is placed on the quality of the thread surface, the absence of scratches and cracks. These damages are often ignored in the dimensional inspection, therefore an important condition for the approval of parts before shipment to the customer is visual inspection, which ultimately eliminates components containing surface defects.

In most cases, in order to check the correctness of the thread, it is sufficient to control it using dedicated gauges: go and no-go. The main advantage of this solution is the speed of verification, low operator skill requirements and the ability to check the functionality of the entire thread at once [9]. Unfortunately, this method has many disadvantages. First of all, the test gives only a binary result and in the case of detecting an irregularity, we are not able to identify which feature it relates to. Moreover, Kosarevsky [10] noticed that it is easy to overlook the incorrectness of the thread if despite the dimensional deviation it meets the functionality, e.g. the pitch error may be hidden by making a larger pitch diameter. In the same work, he also pointed out other disadvantages of the gauges, such as high cost of production and susceptibility to wear, which necessitates regular calibration. Suliga [11] noticed that, based only on checking by gauge, it is possible to ignore defects which in some situations may be critical, e.g. cracks.

Traditional measuring methods have been clearly described by Malinowski [3] in the handbook dedicated to the thread measurements. The flank angle and the root radius could be measured by optical technique using a microscope or a projector. One of the helpful solutions in the case of small dimensions or invisible, hard-to-reach threads is to recreate its profile on the basis of replication [12]. For measuring geometric elements of the thread profile, it can be also using a contourograph equipped with a thin and stiff tip with a small radius (approx. 0.01 mm). However, the author emphasizes, that in this way only information about the total thread angle α is obtained, not the angles of the individual flanks α_1 and α_2. In difference to the optical technique, this contact device it is not able to measure the contour on both sides of the axis. The value of the pitch diameter can be evaluated using a microscope or a projector, as well as by touching method with a micrometer with a special, interchangeable tips with a conical and prismatic shape selected depending on the nominal thread pitch. However, the most accurate in this case will be an indirect method called the three-wire method. Properly selected measuring wires are placed in the thread groove and their outer diameter is measured perpendicular to the thread axis. The relationship between the direct measurement M and the value of the pitch diameter d_2 is given by the formula (1) [3].

$$d_2 = M - d_w \left[1 + \frac{1}{\sin\left(\frac{\alpha}{2}\right)}\right] + \frac{P}{2} ctg\left(\frac{\alpha}{2}\right) + p_1 + p_2 \qquad (1)$$

where: d_w – wire diameter, α – thread angle, P – pitch, p_1 – rake correction, p_2 – correction due to deformation caused by measurement force.

For a better illustration, this relationship is shown in the below figure (Fig. 1).

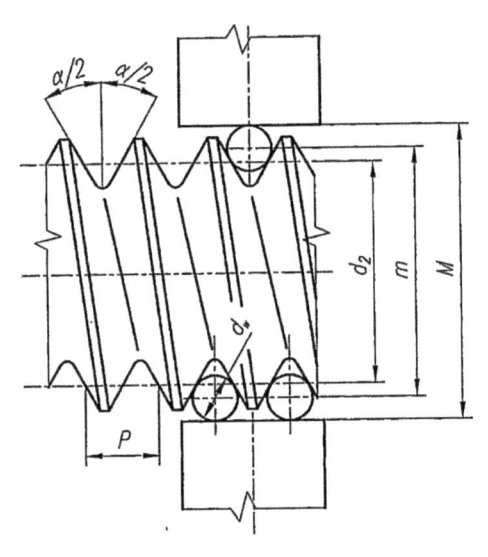

Fig. 1. Three wire method for pitch diameter measurement [3].

The major diameter of the external thread could be measured with the same devices that are used to measure the external diameters of the shafts. In the case of the minor diameter, both optical and contact measurements are used, e.g. with a thread micrometer or a projector.

The continuous development of a coordinate measuring technique made it possible to measure selected thread features at CMM stations. Despite the high price of the machine, its universalism causes that more and more companies decide to use these devices in measuring centers [13]. It is worth to notice that the expensive gauge is dedicated to a specific, one thread, while the measuring machine has no such limitations and is much more universal. The measurements of threads on the CMM are primarily limited by the size of the ball, which is often unable to reach the bottom of the groove [11]. Carmignato [14] performed the test using a special needle-shaped tip with a radius less than 0.1 mm (see Fig. 2). These tests resulted in obtaining the uncertainty of measuring the pitch diameter at the level of 2 μm.

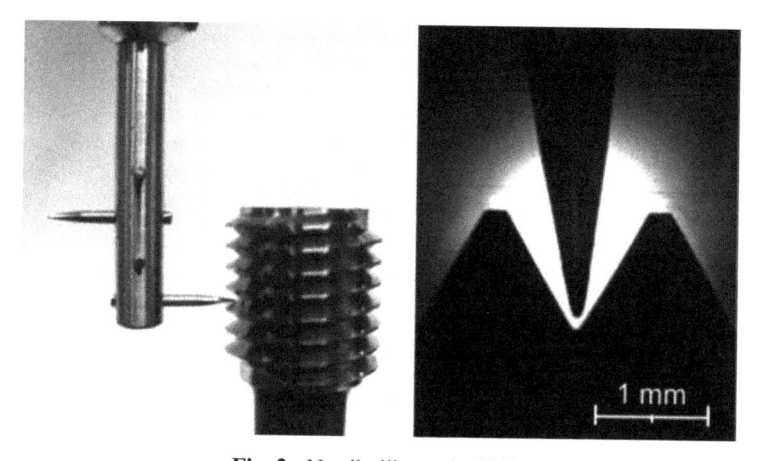

Fig. 2. Needle-like probe [14].

When measuring the threads on the CMM, specific dimensions should be indicated, and what is more, the measurement takes place only in selected sections, which makes it possible to omit a defect that affects the functionality. In that case the advantage of gauges are highlighted, however, in the case of an assembly problem, the gauge control will not give us information where the problem exist, the CMM does [9]. Preparation a program for CMM is certainly time-consuming and requires a qualified operator, but subsequent measurements are made automatically, which eliminates the human factor that occurs with manual methods [15].

Despite the many advantages of CMM, when analyzing the latest literature, a conclusion is drawn that more and more attention and research concerns the non-contact measurement of threads. The arguments for them are speed, the ability to measure many features at the same time, and the ease of collecting and storing data in a digital form. By choosing the non-contact method, the possibility of scratching the thread surface is eliminated due to its non-invasive character. One of such methods, based on laser triangulation, was described by Suliga [11]. The researcher emphasized that the measurement accuracy is limited by the resolution of the light sensor, but increasing the resolution affects the time of data processing. In his experiment, Suliga drew attention to the occlusion effect, i.e. the occurrence of blind spots that are not registered by the measuring system. In the work, he pointed out the advantages of the rotational motion of the measured detail in relation to the measuring system in relation to the classic approach of linear displacement. Light reflections, depending on the type and quality of the measured surface, and the light beam scattering at the edges are a great difficulty in optical measurements [8, 16].

In recent years, there has been a rapid development of information technologies supported more and more often by artificial intelligence. Due to the development of image processing algorithms, vision systems regularly support quality control processes. The measuring system developed by Gadelmawla [17] allows the evaluation of as many as 18 thread features with an accuracy deviating from standard measuring methods by ± 5.4 µm. He, Zhang et al. [18] proposed a method for assessing the extent of damage as well as for measuring of damaged tapered threads based on the linear array CCD. Chen et al. [19] noticed that with non-contact detection of thread profile boundaries, the challenge remains to ensure that the thread axis intersects the CCD camera axis perpendicular. Researchers have described an algorithm that can reduce the workpiece inclination error by 50%. The literature also describes attempts to introduce a measuring system based on vision systems directly into the machining space [20]. Unfortunately, at present, a big problem that makes such a solution practically impossible on an industrial scale is the influence of the lubricating oil and cooling liquid on the obtained measurement results, and it depends on the viscosity of the oil as well as the measured feature. The greatest error was observed when measuring the minor diameter of the external thread [21]. The occurrence of chips and temperature fluctuations are also factors that make it difficult to use an automatic measuring system in the processing environment.

Currently, in industrial factories, devices aimed at improving and accelerating measuring processes are more and more often found. An example can be a series of optical scanners dedicated to shafts and other rotating elements - Opticline produced by the Jenoptik company (Fig. 3). The measurements are based on cameras with high resolution 0.1 µm and light sources moving along the axis of rotation of the workpiece, which guarantees the rapid collection of the contour and the calculation of many measurement features. Dedicated software provides the ability to measure both the features of geometric dimensions and deviations of shape and position, where one of the options is also thread profile measurements. It should be noted that these measurements certainly depend on both external factors and the quality of the device itself, but also the software and image analysis algorithms play an extremely important role here.

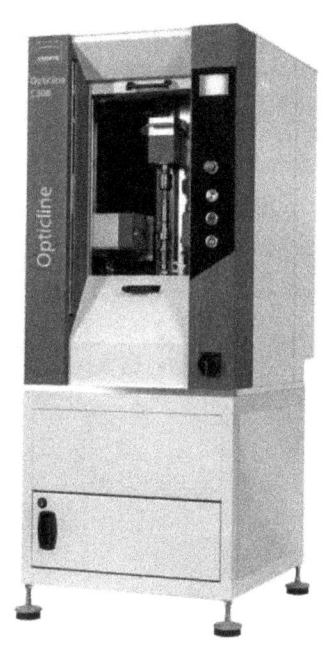

Fig. 3. The measuring station designed for rotating parts measurement [22].

The authors of this study decided to carry out a comparative test of traditional manual methods with the selected method of automatic measurement.

3 Material and Methods

The research was carried out on three test pieces with an external, inch threads. Each of the samples had a different thread profile, triangular (UNJ), trapezoidal symmetrical (ACME) and trapezoidal asymmetrical (BUTTRESS). The dimensional tolerances of the tested elements were specified on the basis of the American standards [23–25]. Nominal dimensions are presented in the table (Table 1).

Two features were measured on each thread, the major diameter and the pitch diameter. In the first step, manual measurements were made. For the external diameter, for

Table 1. Nominal dimensions of tested threads.

	2.750-16 UNJ-3A	2.875-16 ACME-3A	4.5590-12 BUTT-3A
Number threads per inch	16	16	12
Major diameter [mm]	Ø69.73$^{\pm 0.11}$	Ø72.96$^{\pm 0.08}$	Ø115.8$^{\pm 0.07}$
Pitch diameter [mm]	Ø68,767$^{\pm 0.05}$	Ø71,878$^{\pm 0.08}$	Ø114.64$^{\pm 0.07}$
Wire [mm]/M_p measurement [mm]	Ø0.914 Ø70.14$^{\pm 0.05}$	Ø0.914 Ø73.373$^{\pm 0.08}$	Ø1.093 Ø116.12$^{\pm 0.07}$

UNJ and ACME threads, a Mitutoyo digital micrometer with a measuring range of 50–75 mm (No.: 293-232) was used, while for the BUTTRESS thread, a device of the same manufacturer with a range of 100–125 mm (No.: 293-250). The pitch diameters were measured using the three-wire method (Fig. 4). Wires with a diameter of Ø0.914 mm (UNJ, AMCE thread) and Ø1.093 mm (BUTTRESS) were used.

Fig. 4. The view of wires in roots of ACME thread.

The indirect measurement was evaluated with a passameter by setting the recalculated nominal dimension on the stack of gauge blocks (Fig. 5). Mitutoyo 50–75 mm (No.: 523-123) for UNJ and ACME and Steinmeyer 100–125 mm (No.: 276016) for BUTTRESS dial snap meters were used in this experiment. Measurements were made three times for each element in places distributed on the circumference of the parts every 60°. Each of the above micrometers and passameters has a step of 1 µm.

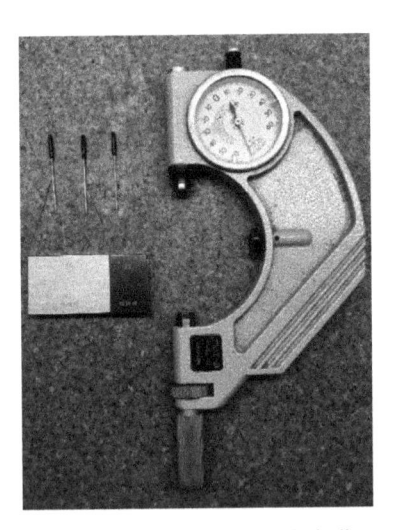

Fig. 5. Measurement devices used to measure pitch diameter ACME thread.

For comparison, the same features were measured with the use of automatic devices. Symmetrical threads were measured optically with Opticline C1214 (Fig. 6). Maximum permissible error – MPE, for diameter 1.0 + D/200 μm and for length 2.6 + L/200 μm. Dimensions was measured in ten sections, evenly spaced around the circumference. The average values are given as a result. For an unsymmetrical trapezoidal thread, the pitch diameter cannot be calculated due to the strong obscuring of the sides of the thread profile in the profile image. For this reason, the buttress thread was measured by a different method. For this purpose, Zeiss ACCURA II 12/30/10 coordinate measuring machine with length measurement error 1.9 + L/300 μm was used. The measurement was performed with the touching probe with the ruby stylus with diameter Ø1 mm (Fig. 7). The pitch diameter was measured by scanning the groove along the full circumference of the shaft in one section. The self-centering probing function available in the Calypso 6.2 software was used [15]. The major diameter was also checked with a full circumference crest scan. In both cases, the average values were evaluated.

Fig. 6. The view of ACME thread measurement on Opticline C1214.

Fig. 7. The view of buttress thread measurement on CMM.

4 Results and Discussion

The results of the measurements are presented in the table below (Table 2).

Table 2. Measured dimensions of tested threads.

		2.750-16 UNJ-3A	2.875-16 ACME-3A	4.5590-12 BUTT-3A
Major diameter [mm] –micrometer	Avg.	Ø69.753	Ø72.957	Ø115.793
	Min.	Ø69.752	Ø72.955	Ø115.784
	max.	Ø69.753	Ø72.961	Ø115.805
Major diameter [mm] – Opticline	Avg.	Ø69.756	Ø72.954	
	Min.	Ø69.754	Ø72.951	–
	Max.	Ø69.757	Ø72.959	
Major diameter [mm] – CMM	Avg.	–	–	Ø115.796
	Min.			Ø115.781
	Max.			Ø115.816
Pitch diameter [mm] – three wires	Avg.	Ø68.773	Ø71.866	Ø114.632
	Min.	Ø68.771	Ø71.863	Ø114.626
	Max.	Ø68.775	Ø71.870	Ø114.640
Pitch diameter [mm] – Opticline	Avg.	Ø68.786	Ø71.877	
	Min.	Ø68.783	Ø71.873	–
	Max.	Ø68.788	Ø71.882	
Pitch diameter [mm] – CMM	Avg.	–	–	Ø114.647
	Min.			Ø114.631
	Max.			Ø114.655

Each of threads has been made in accordance with the dimensional tolerance. When comparing the average results of measuring major diameters for three types of threads, it was noticed that the maximum difference between the manual and automatic methods is 3 μm, which confirms the validity of the interchangeability of both methods. In the case of comparative studies on pitch diameters, both for Opticline and CMM, greater differences were obtained, reaching a maximum of 15 μm. This measurement is an indirect measurement, certainly with a greater error. Moreover, the manual measurement was limited to three results, while the average results obtained automatically included many more points. It has been noticed that the applied optical method also does not ensure full accessibility to the thread profile in the section dividing shaft along the axis because of the obscuring. In the case of the optical method, the cause can also be found in the software itself and in the algorithm for calculating the pitch diameter value. When analyzing the minimum and maximum values of the buttress thread made of Inconel 718, it was noticed that the diameter had an oval of 35 μm, which is a common phenomenon for this material. Certainly, the cleanliness of the surface, the calibration of devices, as well as the determination of the base surface in the case of CMM, have an influence on the measurement results. To sum up, the tested methods are suitable for measuring details in industrial production, but in the case of narrow tolerances, the selected method should certainly verify and their correctness should be confirmed, e.g. by the third method. Due to the many advantages, the requirements of the modern manufacturing industry mean that measurements should be performed automatically.

Further research into automatic thread measurement methods are planned. In the next stage, the tests carried out will be extended to a larger number of samples and measurement methods like e.g. laser line [26]. Tests of measuring the pitch diameter will also be performed directly on a CNC machine e.g. with a Renishaw RMP600 measuring probe.

5 Conclusions

1. Quality assurance of threaded elements is an extremely important and complex task, especially if they are used in aviation applications.
2. Research on the automation of the measurement of external threads is described in the literature to a very limited extent, especially when it comes to unsymmetrical trapezoidal threads.
3. Automatic measurements based on optical technology are faster and more universal than contact measurements, but they are more susceptible to external factors, such as cleanliness or light reflections.
4. On the basis of the obtained results, it can be confirmed that it is possible to automate the measurement of selected features of external threads.
5. According to the literature, attempts are being made to automate the measurement of external threads on CNC machines, but currently there is no ready-made solution used on an industrial scale.

Acknowledgments. The authors thank the Polish Ministry of Education and Science for financial support (Applied Doctorate Program, No. DWD/4/22/2020).

This research was funded by the Polish Ministry of Higher Education grants no. 0614/SBAD/1547.

References

1. Tong, Q.-B., Han, B.-Z., Wang, D., Wang, J.-Q., Ding, Z.-L., Yuan, F.: A novel laser-based system for measuring internal thread parameters. J. Russ. Laser Res. **35**(3), 307–316 (2014). https://doi.org/10.1007/s10946-014-9429-0
2. Hong, E., Zhang, H., Katz, R., Agapiou, J.S.: Non-contact inspection of internal threads of machined parts. Int. J. Adv. Manuf. Technol. **62**, 221–229 (2012). https://doi.org/10.1007/s00170-011-3793-5
3. Malinowski, J., Jakubiec, W., Plowucha, W.: Pomiary gwintow w budowie maszyn. Wydawnictwo Naukowo-Techniczne, 2nd edn., Warszawa 2008 (2010). ISBN 978-83-204-3626-6

4. Grilli, M.L., et al.: Critical raw materials saving by protective coatings under extreme conditions: a review of last trends in alloys and coatings for aerospace engine applications. Materials **14**, 1656 (2021). https://doi.org/10.3390/ma14071656
5. Khanna, N., Agrawal, C., Dogra, M., Pruncu, C.I.: Evaluation of tool wear, energy consumption, and surface roughness during turning of Inconel 718 using sustainable machining technique. J. Matter. Res. Techol. **9**(3), 5794–5804 (2020). https://doi.org/10.1016/j.jmrt.2020.03.104
6. He, Y., Zhou, Z., Zou, P., Gao, X., Ehmann, K.F.: Study of ultrasonic vibration–assisted thread turning of Inconel 718 superalloy. Adv. Mech. Eng. **11**(10), 1–12 (2019). https://doi.org/10.1177/1687814019883772
7. Fong, K.M., Wang, X., Kamaruddin, S., Ismadi, M.Z.: Investigation on universal tool wear measurement technique using image-based cross-correlation analysis. Measurement **169**(108489), 1–13 (2021). https://doi.org/10.1016/j.measurement.2020.108489
8. Czajka, P., Garbacz, P., Mężyk, J., Giesko, T., Mazurkiewicz, A.: The optomechatronics system for automatic quality inspection of machined workpieces. J. Mach. Constr. Maintenance **4/2018**(111), 77–86 (2018). ISSN 1232-9312
9. Hill, C.: CMMs vs. thread gages. Qual. Mag. **55**, 16 (2016)
10. Kosarevsky, S., Latypov, V.: Detection of screw threads in computer tomography 3D density fields. Meas. Sci. Rev. **13**(6) (2013). https://doi.org/10.2478/msr-2013-0043
11. Suliga, P.: A feature analysis of a laser triangulation stand used to acquire a 3D screw thread image. In: 17th International Carpathian Control Conference, pp. 702–705 (2016)
12. Marteau, J., Wieczorowski, M., Xia, Y., Bigerelle, M.: Multiscale assessment of the accuracy of surface replication. Surf. Topogr. Metrol. Prop. **2**(4) (2014). Art. no. e044002. https://doi.org/10.1088/2051-672X/2/4/044002
13. Gapinski, B., Wieczorowski, M., Marciniak-Podsadna, L., Dybala, B., Ziolkowski, G.: Comparison of different method of measurement geometry using CMM, optical scanner and computed tomography 3D. Procedia Eng. **69**, 255–262 (2014). https://doi.org/10.1016/j.proeng.2014.02.230
14. Carmignato, S., De Chiffre, L.: A new method for thread calibration on coordinate measuring machines. CIRP Ann. **52**(1), 447–450 (2003). https://doi.org/10.1016/S0007-8506(07)60622-2
15. Yüksel, I.A., et al.: Comparison of internal and external threads pitch diameter measurement by using conventional methods and CMM's. In: 19th International Congress of Metrology, 09001 (2019). https://doi.org/10.1051/metrology/201909001
16. Gapiński, B., et al.: Use of white light and laser 3D scanners for measurement of mesoscale surface asperities. In: Diering, M., Wieczorowski, M., Brown, C.A. (eds.) MANUFACTURING 2019. LNME, pp. 239–256. Springer, Cham (2019). https://doi.org/10.1007/978-3-030-18682-1_19
17. Gadelmawla, E.S.: Computer vision algorithms for measurement and inspection of external screw threads. Measurement **100**, 36–49 (2017). https://doi.org/10.1016/j.measurement.2016.12.034
18. He, F.J., Zhang, R.J., Du, Z.J., Cui, X.M.: Non-contact measurement of dam-aged external tapered thread based on linear array CCD. J. Phys. Conf. Ser. **48**, 676–680 (2006). https://doi.org/10.1088/1742-6596/48/1/127
19. Chen, J.H., Zhang, J.J., Gao, R.J., et al.: Research on modified algorithms of cylindrical external thread profile based on machine vision. Meas. Sci. Rev. **20**(1), 15–21 (2020). https://doi.org/10.2478/msr-2020-0003
20. Lee, Y.C., Yeh, S.S.: Using Machine vision to develop an on-machine thread measurement system for computer numerical control lathe machines. In: IMECS 2019, 13–15 March, Hong Kong (2019)

21. Yang, Z., Chen, M., Wang, P.: Effect of oil adhesion on the measurement of screw thread with machine vision. Meas. Control **53**(5–6), 922–933 (2020). https://doi.org/10.1177/0020294020903477
22. https://www.ita-polska.com.pl/. Accessed 27 Oct 2021
23. ASME B1.1-2003 (R2018)
24. ASME B1.5-1997 (R2009)
25. ANSI B1.9-1973
26. Swojak, N., Wieczorowski, M., Jakubowicz, M.: Assessment of selected metrological properties of laser triangulation sensors. Meas. J. Int. Meas. Confederation **176** (2021). Art. no. 109190. https://doi.org/10.1016/j.measurement.2021.109190

New Methodology of Face Mill Path Correction Based on Machined Surface Measurement to Improve Flatness

Marek Rybicki

Poznan University of Technology, ul. Piotrowo 3, 60-965 Poznan, Poland
marek.rybicki@put.poznan.pl

Abstract. In the article is presented research results of deviations in various longitudinal sections (measured parallel to feed direction) and deviations in transverse sections (measured perpendicular to feed direction) during face finish milling. It was found that diversification of deviations in various sections is a result of geometrical action of the mill relative to workpiece movement. Original method of improving flatness by compensation path preparing and assigning the path to face mill was presented. Compensation was realized in interrupted way, consisting in determination of compensation path on the basis of machined surface profiles measurement in the pass before compensation one. It has been proposed to carry out pass with tilted axis of the mill before compensation to avoid back cutting and obtain correct compensation path. Then in compensation pass axis of the mill should be set perpendicular and compensation path should be assigned to front part of the mill, which formed machined surface in previous pass. Face mill cuts on length higher than length of workpiece. So it has been proposed to take not only single longitudinal profile but also fragment of longitudinal profile distanced the most from axis of the mill as a compensation path shape. The developed compensation technique is more advantageous in relation to applied to date ones, during milling with large diameter of mills and for large longitudinal deviations.

Keywords: Face milling · Flatness · Compensation · Tool path correction

1 Introduction

Since machined surfaces of many workpieces, e.g. deck faces of engine blocks or mould joints, meets role of sealing faces, flatness of the faces is of great practical importance [1]. Simultaneously from many years is visible tendency of replacing some grinding operations with precision turning and milling.

Methods of machined surface flatness error assessment are often based on registration of profiles in consecutive sections parallel or perpendicular to feed direction (deviations Δw in longitudinal sections or deviations Δp in transverse sections respectively) [2]. This is justified by different arising mechanisms of deviations presenting in the two directions. The longitudinal deviation arises mainly from workpiece and wedge expansion due to cutting heat [3, 4], regret of wedge corner due to its wear [4], elastic deformation of machined system due to cutting forces [5–12] and geometric errors of the system [7, 11, 13]. As a main reason for the transverse deviations arising is mentioned tilt of face mill axis [1, 8].

In face milling process deviations Δw in various longitudinal sections and deviations Δp in various transverse sections are diversified. Literature sources indicate that the deviations can have various run. Reasons for the diversification are recognized weekly. Common feature of research results of [1, 6] concerning modeling of cutting forces impact on flatness error after face milling by the FEA method, is the surface rising at face mill exit in right corner of the surface (looking from beginning of milling length). Authors of [2] investigated real deviations after face milling in conditions assumed in modeling in [1]. They obtained the lowest profiles in transverse sections distanced from clamping elements of workpiece, inversely than in [1], what they supposed to be result of another face mill and its tilt angle application.

Due to the shape error identification scheme, two types of compensation can be distinguished: "off-line" (passive) and "on-line" (active). In off-line compensation, errors are first measured or identified, and then corrected in successive passes. In off-line compensation, the identification of errors and their control are carried out simultaneously during machining.

The error profile shape and its value can be identified in the "off-line" method: before machining process, after the process or intermittently. With pre-process identification, a model of the influence of various factors on formation of the errors is first created, which is then used to increase accuracy during machining. In after process method errors are identified by measurement. The idea of an intermittent process is to measure workpiece errors on the machine after replacing cutting tool with a measuring tool. In the "on-line" method, two main approaches to identifying of errors can be distinguished: registration of errors during cutting or their identification using mathematical prediction.

The use of the mathematical model during face milling is used when one or two sources of errors, e.g. cutting forces [7–9, 11–15] or geometric errors of the machine tool [7, 11, 13] dominate the others. The advantage of identifying errors on the basis of measurements of the machined surface [13, 16] is that all error sources are taken into account, except for the random errors that can only be taken into account in active compensation.

After identifying errors, compensation should be performed. There are three main methods of its implementation in face milling: by means of variable feed along the milling length [14, 16–18], deformation of the workpiece [3, 10] and correction of the tool path [7, 9, 12–14, 16, 18]. The authors [14, 16, 18] proposed compensation by combination of the variable feed strategy and modification of the vertical position of the cutter.

Each of the compensation methods has some disadvantages. The use of variable feedrate to achieve a specific face mill deformation affecting form errors is difficult to implement and the errors reduction is limited. In turn, the use of deformation of workpiece by clamping forces is only suitable for improving flatness of workpieces with selected shapes. During compensation by path correction method, the main problem is back cutting by face mill [13, 14, 16, 18]. Authors of [13] compensated deviation in longitudinal section during face milling by tool path assigned to axis of small diameter end mill. In the authors opinion assigning of compensation path to axis of large diameter face mills could cause "undercutting" by back of the mill problem, especially when component of longitudinal deviation from geometrical errors of machine tool is large. In order to otain flat surface after face milling it is necessary to set axis of the mill perpendicular to feed direction, when bi-directional lay on machined surface occur. At Fig. 1 is visible that when axis of the mill is perpendicular to feedrate, another profile of machined surface is obtained than the path shape during movement of the mill according to curvilinear path, due to back cutting effect.

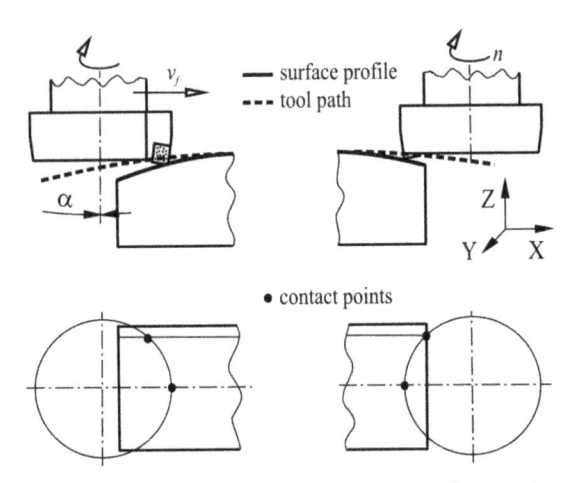

Fig. 1. Contact of wedge in various points of machined surface during face milling with curvilinear path and non-tilted axis of the mill.

2 Range and Conditions of Research

The research was divided into three parts. Description of phenomena connected with arising of machined surface deviations in various longitudinal and transverse sections is presented in part 3.1 of the work. Minimization of the deviations by correction of face mill path is presented in part 3.2 and results analysis of the compensation in part. 3.3.

Research we carried out during symmetrical dry face milling of hot work tool steel 55NiCrMoV in hardened state. Constant cutting parameters v_c, f_z, a_p and constant ratio of milling width to mill diameter $a_e/d = 0,9$ were applied in each research parts (Table 1). There was applied fly milling, for the sake of possible difficulties in results interpretations in conditions of wedges run out. Passes in the deviations diversity research and compensation tests were carried out by brand new wedges. Wedge wear characterized by band width of the corner wear VB_c was changing in range given in Table 1.

Table 1. Selected machining conditions

Machine tool:	FND32F made of AVIA (in part 3.1), FYN50ND made of JAFO (in part 3.3)
Workpiece:	55NiCrMoV (52HRC), $a_e = 78,5$ mm, $l = 215$ mm (in part 3.1), $l = 250$ mm (in part 3.3),
Face mill:	4.01006R232 made of Kennametal $d = 87$ mm, $z = 1$, $\alpha_p = 6°$, $\gamma_p = -6°$, $\gamma_f = -11°$
Machining parameters:	$v_c = 306$ m/min, $a_p = 0,15$ mm, $f_z = 0,125$ mm/wedge
Cutting insert:	RNGN 120700 T01020, KY4300 (Al_2O_3 + SiC) grade made of Kennametal
Wedge wear:	$VB_c = 0$–$0,055$ mm

For registration in time of machined surface profiles incremental length gauge MT12B made of Heidenhain with resolution 0,0005 mm and own measuring program was used (Fig. 2). The gauge was fixed to frame of milling machine and it was moving during measurement in axes X and Y of the machine and workpiece coordinate system. It was assumed that during registration of the profiles on milling machine geometrical errors of its movable units not superimposed to the profiles (it was found that milling machine geometrical errors impact was less than 2 μm [9]). The deviations values Δw and Δp were gave as a distance between maximal and minimal position of approximated profiles $Z = f(X)$ and $Z = f(Y)$ measured relative to movement path of the gauge.

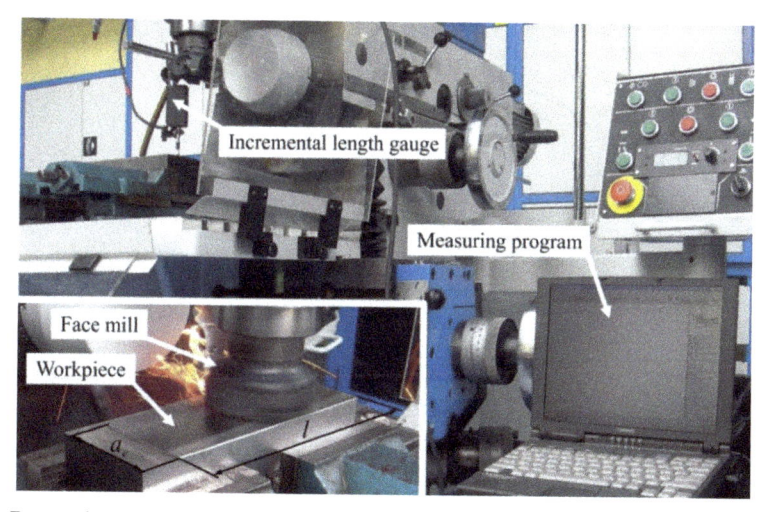

Fig. 2. Research stand for face milling and registering of machined surface profiles at FND32F milling machine.

3 Results and Analysis of Research

3.1 Deviations Diversity in Various Sections

Understanding of deviations arising in various sections is necessary to carry out the compensation. The deviations were analyzed parallel and perpendicular to feed direction due to different factors that have impact on their arising.

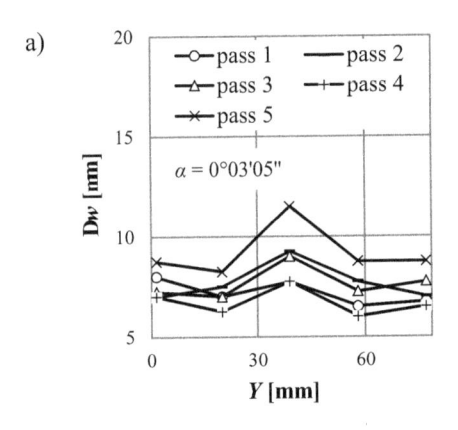

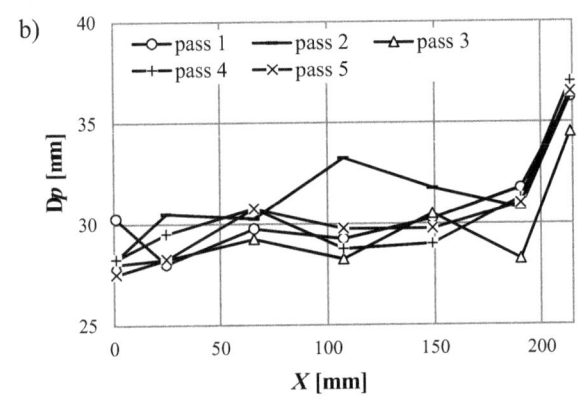

Fig. 3. Deviations in various longitudinal a) and transverse b) sections (pass 1–5 is the repeating of milling and measurements).

In middle longitudinal section intersecting axis of face mill deviation Δw had the highest value (Fig. 3). The phenomenon can be explained by delay in generation of machined surface by wedge of face mill in side longitudinal sections relative to middle section. From the moment the wedge was begun to cut work material in middle section II to the moment cutting has begun in side sections I and III, distance l_ψ passes (Fig. 4). During applied symmetrical milling the l_ψ length can be calculated from Eq. 1. During the l_ψ distance has been running deviation $\Delta w(l_\psi)$ was arisen in section II. Because only just from the moment cutting has begun in sections I and III, deviations in the sections will can attain fewer values than in section II.

$$l_\psi = 0,5\left(d - \sqrt{d^2 - 4b^2}\right) \tag{1}$$

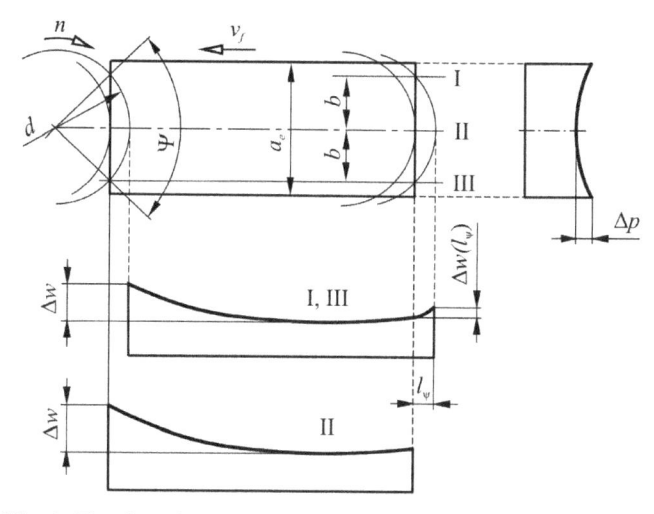

Fig. 4. Forming of machined surface in various longitudinal sections.

As it was mentioned course of transverse deviations can be various on milling length. The course can be explained by variability of face mill axis tilt relative to real path of the mill. When feed direction is parallel to the X axis deviation component $\Delta p(\alpha)$ caused by tilt of face mill axis by α angle, can be calculated during symmetrical milling from relationship (2) But tilt of face mill axis in the XZ plane changes by $d\alpha$ value in relation to tangent to the profile $Z = f(X)$ in longitudinal section due to the profile curvature (Eq. 3, Fig. 5). Assuming that $l_\psi = 0$ mm deviation value $\Delta p(\alpha + d\alpha)$ caused by the curvature can be written down by formula 7.

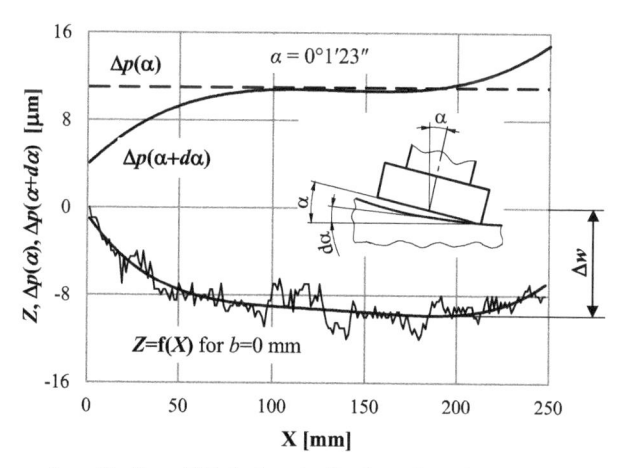

Fig. 5. Influence of profile $Z = f(X)$ in longitudinal section shape on deviation in transverse section component $\Delta p(\alpha + d\alpha)$ caused by real tilt of face mill axis.

$$\Delta p(\alpha) = 0,5\left(d - \sqrt{d^2 - a_e^2}\right)\sin\alpha \tag{2}$$

$$d\alpha = \tan^{-1}\frac{dZ}{dX} \tag{3}$$

$$\Delta p(\alpha + d\alpha) = 0,5\left(d - \sqrt{d^2 - a_e^2}\right)\sin(\alpha + d\alpha) \tag{4}$$

Theoretical value of deviation $\Delta p(\alpha + d\alpha)$ on milling length is presented against the background of real longitudinal profile $Z = f(X)$ in Fig. 5. A bit another real course $\Delta p = f(X)$ (Fig. 3b) than course $\Delta p(\alpha + d\alpha) = f(X)$ at Fig. 5 can be explained by the l_ψ delay. At the figure is given the α angle, measured by incremental length gauge. Tilt by α angle was applied intentionally to give unidirectional lay on whole machined surface, because it was easier case for analysis of the deviations arising.

3.2 Compensation Method

There was analyzed passive compensation, with identification of errors in previous pass and their correction in next one. Identification of the errors was performed intermittently, consisting in performing milling, then replacing the cutting tool with a measuring tool. Then, after developing the compensation path, a milling pass was made with the same parameters and tool path correction. Measurement of machined surface on the machine tool in the intermittent method takes into account all error influencing factors, except the machine geometric errors. It is effective in serial production of the same parts due to time consuming procedure of compensation path determination before right machining.

It was found that in pass before compensation axis of the mill has to be tilted in feed direction by angle $\alpha > 0°$ (1 in Fig. 6), to be sure back of the mill will not be cut again surface formed by its front. Only then there is certainty the movement path of the mill will corresponds to machined surface profile. Axis of the mill has to be set perpendicular to feed direction in compensation pass only 8. Setting of the axis relative to feed direction should be done at the beginning of milling length for $l = 0$ for the two passes, because real feed direction is changeable and causes variability of the tilt on length l by $d\alpha$ angle (Eq. 4, Fig. 5). For the sake of various tilt settings necessity in the passes before and during compensation the presented procedure cannot be applied in active compensation.

After pass before compensation 2 longitudinal profiles in axis of the mill and distanced by b from axis of the mill should be registered 3. It needed for creation of compensations path shape by connecting of registered profiles fragments (Fig. 7).

On milling length segment $(0; l)$ the path shape should be assumed to be longitudinal profile crossing axis of the mill (or placed the nearest to the axis if no profile crossing it during non-symmetrical milling) 4. Face mill cuts at length higher than workpiece length l by the l_ψ value. If compensation path shape would be assumed to be longitudinal profile crossing axis of the mill only, area in exit phase of the mill at length $(l; l + l_\psi)$ will not be compensated. So at the $(l; l + l_\psi)$ length compensation path can be fragment of longitudinal profile registered at distance b from axis of the mill 5. It can be done because geometrically every point of the mill moves with the same path.

To obtain compensation path created shape should be reflected relative to nominal machined surface profile 6. Beginning of compensation path should be assigned to point on front part of the mill, which cut longitudinal profile the nearest to axis of the mill 7, because front part of the mill cut machined surface in pass 2 before compensation. Assigning of compensation path to other point of the mill will caused shift of the path on milling length.

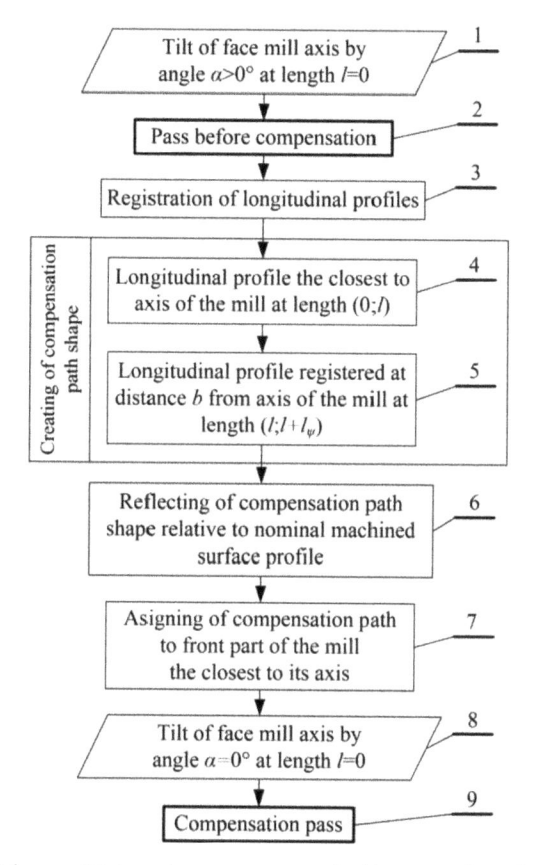

Fig. 6. Scheme of flatness improvement method by correction of mill path.

Over length $l + l_\psi$ the mill can still cut by its back. If geometrical errors were not earlier inputted to control system of machine tool, deviation component from the geometrical errors should be took into account over the length, because other factors are equal to zero.

Giving consideration only to impact of geometrical movement of the mill it can be stated that other versions of longitudinal profiles fragments joining to determine compensation path shape than presented in Fig. 6 and 7 are possible. For instance, compensation path shape can be equal to profile registered the closest to axis of the mill on milling length (0; l_ψ) and profile the farthest from the axis on length (l_ψ; $l + l_\psi$).

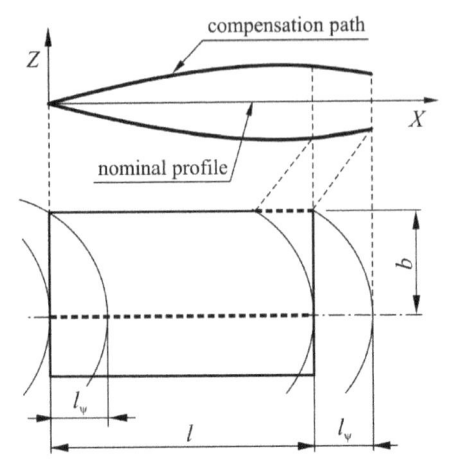

Fig. 7. Creating of compensation path shape.

3.3 The Compensation Effects

In machining tests beginning of compensation path was assigned to point in front of the mill crossing axis of the mill, like at Fig. 7. Compensation path shape was assumed to be longitudinal profile registered in axis of the mill on length (0; l) and profile distanced by $b = 36$ mm from the axis on length (l; $l + l_\psi$). Tilt of face mill axis by angle $\alpha > 0$ in pass before compensation and by angle $\alpha = 0$ in compensation pass were applied.

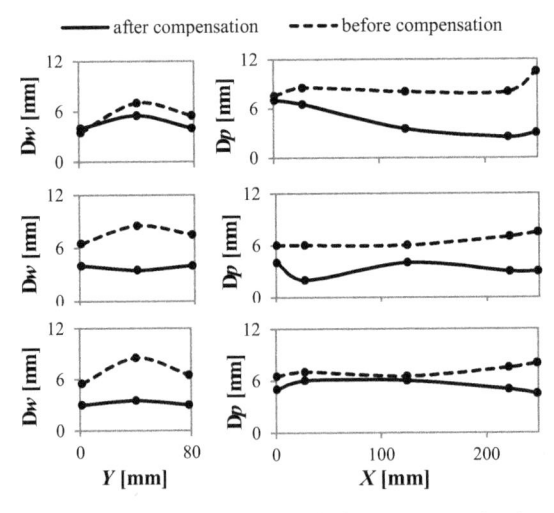

Fig. 8. Deviations Δw and Δp obtained before and after compensation in various longitudinal and transverse sections for three repeating of the experiment.

Figure 8 shows the results of the deviations Δw and Δp in different sections for the three compensation tests performed. Correction of face mill path in relation to the workpiece allowed to reduce to about 5 μm and to equalize values of deviations Δw in all longitudinal sections. It is probably limited capability of machine tool used. It can be assumed that flatness improvement takes place when decrease not only longitudinal deviations but also transverse ones. After compensation transverse deviations also decreased, but this is mainly result of reduction of face mill axis tilt angle α.

Figure 9 shows the longitudinal machined surface profiles recorded in three different sections before and after compensation, and the compensation path for one of the performed tests. At the figure is pointed out deviation $\Delta w(l_\psi)$, which will not be compensated if vertical position of the mill would be corrected at length l only.

The error component $\Delta w(l_\psi)$ could be large especially in the case of high longitudinal deviations, large diameter face mills and milling widths.

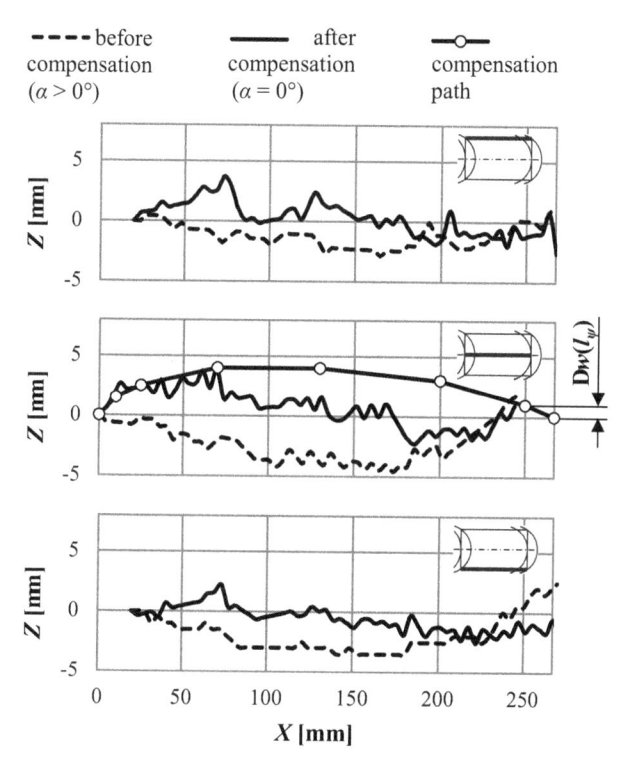

Fig. 9. Influence of compensation path on machined surface profiles in various longitudinal sections.

The smallest reduction of transverse deviations after compensation occurred at the beginning of the milling length. It is caused by rising compensation path, what can be proved by performing milling with a constant inclination angle $\alpha > 0$ in both passes, before and during compensation (Fig. 10). At the figure is visible that rising of compensation path by 5 μm on length $X = l_\psi$ caused rising of transverse profile at the ends of workpiece width measured on length X = 1 mm by similar value. It is caused by increasing of the mill axis tilt angle $\alpha + d\alpha$ measured relative to tangent to compensation path in contact point of tool with workpiece.

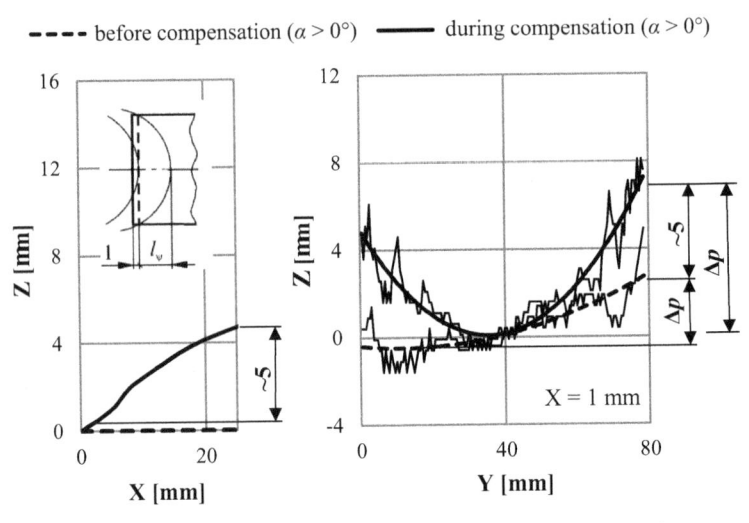

Fig. 10. Influence of face mill path on profile in transverse section.

4 Conclusions

In longitudinal section forming by axis of face mill deviation Δw often has the highest value. It is caused by shift in generation of machined surface by the mill in side longitudinal sections.

Curvilinear profile in longitudinal section, which is a real path of face mill, causes local changes of face mill axis tilt by $d\alpha$ angle. It has effect on deviations in transverse sections Δp variability, measured in various places on milling length.

In order to obtain flat surface after face milling axis of the mill have to be set perpendicular to feed direction. But then face mill path (which is wanted to be compensated) can be different from machined surface profile. So it has been proposed original conception to carry out pass with tilted axis of the mill before compensation to be sure back of the mill will not cut again already machined surface. Then in compensation pass axis of the mill should be set perpendicular and compensation path should be assigned to front part of the mill, which formed machined surface in previous pass.

Face mill cuts on length higher than length l of workpiece. It causes that compensation path shape cannot be equal to single longitudinal profile of machined surface, since some area of the surface will not be compensated. So it has been proposed new strategy to take longitudinal profile in axis of the mill at length $(0; l)$, and to take fragment of longitudinal profile distanced by b from axis of the mill as a compensation path shape at length $(l; l_\psi)$. Beginning of compensation path should be assigned to the same point of the mill, which cut longitudinal profile in axis of the mill in pass before compensation.

The developed compensation technique can be more advantageous in relation to applied to date ones, especially during face milling with large diameter mills, milling widths and longitudinal deviations.

References

1. Gu, F., Melkote, N., Kapoor, S.G., DeVor, R.E.: A model for prediction of surface flatness in face milling. ASME J. Manuf. Sci. Eng. **119**, 476–484 (1997)
2. Badar, A.M., Raman, S., Pulat, S.P.: Experimental verification of manufacturing error pattern and its utilization in form tolerance sampling. Int. J. Mach. Tools Manuf. **45**, 63–73 (2005)
3. Huang, Y., Hoshi, T.: Improvement of flatness error in milling plate-shaped workpiece by application of side-clamping force. J. Int. Soc. Precis. Eng. Nanotechnol. **24**, 364–370 (2000)
4. Rybicki, M., Kawalec, M.: Form deviations of hot work tool steel 55NiCrMoV (52HRC) after face finish milling. Int. J. Mach. Mach. Mater. **7**(3/4), 176–192 (2010)
5. Lan, J.V., Larue, A., Lorong, P., Coffignal, A.: Form error prediction of gearcases' face milling. Int. J. Mater. Form. **1**, 543–546 (2008)
6. Liao, Y.G., Hu, S.J.: An integrated model of a fixture-workpiece system for surface quality prediction. Int. J. Adv. Manuf. Technol. **17**, 810–818 (2001)
7. Habibi, M., Arezoo, B., Nojedeh, M.V.: Tool deflection and geometrical error compensation by tool path modification. Int. J. Mach. Tools Manuf **51**, 439–449 (2011)
8. Pimenov, D.Y., Guzeev, V.I., Mikolajczyk, T., Patra, K.: A study of the influence of processing parameters and tool wear on elastic displacements of the technological system under face milling. Int. J. Adv. Manuf. Technol. **92**(9–12), 4473–4486 (2017). https://doi.org/10.1007/s00170-017-0516-6
9. Tyapin, I., Kaldestad, K.B., Hovland, G.: Off-line path correction of robotic face milling using static tool force and robot stiffness. In: 2015 IEEE/RSJ International Conference on Intelligent Robots and Systems (IROS), 28 September–2 October, Hamburg, Germany, pp. 5506–5511 (2015)
10. Yi, W., Jiang, Z., Shao, W., Han, X., Liu, W.: Error compensation of thin plate-shape part with prebending method in face milling. Chin. J. Mech. Eng. **28**(1), 88–95 (2015). https://doi.org/10.3901/CJME.2014.1120.171
11. Du, Z., Ge, G., Xiao, Y., Feng, X., Yang, J.: Modeling and compensation of comprehensive errors for thin-walled parts machining based on on-machine measurement. Int. J. Adv. Manuf. Technol. **115**(11–12), 3645–3656 (2021). https://doi.org/10.1007/s00170-021-07397-5
12. Checchi, A., Costa, D.G., Merrild, C.H., Bissacco, G., Hansen, H.N.: Offline tool trajectory compensation for cutting forces induced errors in a portable machine tool. Procedia CIRP82, 527–531 (2019)
13. Park, C.W., Eman, K.F., Wu, S.M.: An in-process flatness error measurement and compensatory control system. ASME J. Eng. Ind. **110**, 263–270 (1998)
14. Tai, B.L., Stephenson, D.A., Shih, A.J.: Improvement of surface flatness in face milling based on 3-D holographic laser metrology. Int. J. Mach. Tools Manuf. **51**, 483–490 (2011)
15. Liu, E.A., Zou, Q.: Machined surface error analysis - a face milling approach. J. Adv. Manuf. Syst. **10**(2), 293–307 (2011)
16. Tai, B.L., Stephenson, D.A., Shih, A.J.: Improvement of surface flatness in face milling by varying the tool cutting depth and feed rate. In: ASME 2009 International Manufacturing Science and Engineering Conference, MSEC2009, 4–7 October, West Lafayette, Indiana, USA, vol. 2, pp. 73–79 (2009)

17. Tai, B.L., Wang, H., Nguyen, H., Hu, S.J., Shih A.: Surface variation reduction for face milling based on high-definition metrology. In: Proceedings of the ASME 2012 International Manufacturing Science and Engineering Conference, MSEC2012, 4–8 June 2012, Notre Dame, Indiana, USA, pp. 1–10 (2012)
18. Gu, J., Agapiou, J.S.: Approaches for improving surface flatness for face milling. Trans. North Am. Manuf. Res. Inst. SME **42**, 542–553 (2014)

Development of the Handheld Measuring Probe for a 3D Scanner

Robert Kupiec, Wiktor Harmatys, Izabela Sanetra, Katarzyna Składanowska, and Ksenia Ostrowska(✉)

Laboratory of Coordinate Metrology, Cracow University of Technology, al. Jana Pawła II 37, 31-864 Cracow, Poland
ksenia.ostrowska@pk.edu.pl

Abstract. This paper presents a specially designed contact measuring probe for the 3D scanner. It also presents stages of development of the handheld measuring probe including adjustment on coordinate measuring machine. The test procedure of the system was performed according to VDI/VDE 2634. This procedure involved the determination of probing error, sphere spacing error and flatness measurement error. The performance of the system was verified according to the test procedure. The maximum permissible errors of the system were determined and the measurement uncertainty was taking into account. The accuracy of the system has been improved thanks to the use of the probe calibration on the coordinate measure machine.

Keywords: 3D scanner · Validation · Handheld probes · Markers

1 Introduction

The popularity of three-dimensional 3D scanning technologies have increased over the last few years. This is a result of technological progress for example, scanners are smaller, more portable and more powerful. Furthermore, we can observe increasing range of different applications in multiple industries [1, 2].

The test results described in this article are based on measurements taken by the scanner, such as MICRON3D green stereo (Fig. 1). Depending on the technology we use, there are scanner based the projection of structured light and laser projection. In this research we use structured light scanner with two cameras and one projector. In this scanner, the principle of operation is based on passive triangulation between each camera, projector and measuring point. Geometry mapping is carried out by projecting structured light onto the scanned surface according to the Grey code principle and phase shifted fringes. The MICRON 3D green stereo scanner uses to measure green LED light, with a wavelength of around 500 nm, which enables increase the accuracy of the measurement; even 30% in relation to the traditional method of white light measurement. The use of monochrome cameras enable to create better mapping of the surface structure. In addition, the usage of LED light source reduces energy consumption and increases the service life of the entire system.

All results are based on measurements using the contact probe with determined markers and the 3D scanner. The markers placed on the probe are necessary to determine the position of the probe in the measuring space of the optical scanner - the markers are tracked by the scanner. The handheld probe is used to measure elements most often invisible to the scanner for example internal holes. This contact probe is useful for measuring process in which the scanner may have an obstacle to measure complicated objects [3–9].

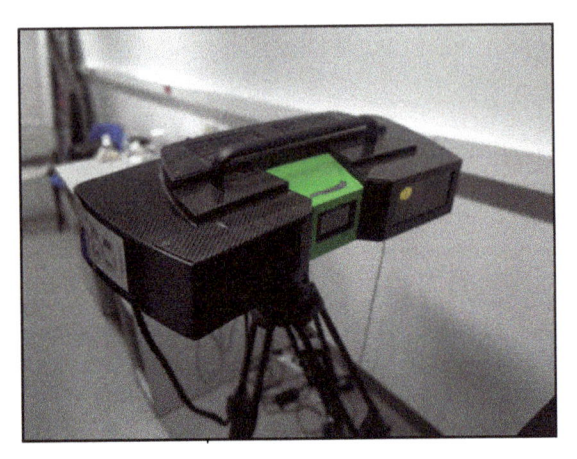

Fig. 1. Scanner MICRON 3D green stereo.

2 Stages of Development of the Tactile Probe

Development of contact probe had several important stages. The first step was the choice of appropriate markers. This part of research was aimed at selecting a material that would ensure the most appropriate representation of the marker in the image made by the scanner detector. Research work was carried out for cameras operating in the visible and infrared beams. The selection of markers was influenced by the shape of the probe and the number of points needed for a reliable identification on the probe in the measurement space. The possibility of their calibration on a more accurate device than the designed system was also taken into account.

At the beginning, tests were carried out with passive markers (Fig. 2), which were in the form of stickers or a printout on the probe. After testing, it turned out that the passive markers were insufficient, the main problem was the inability to precisely calibrate them.

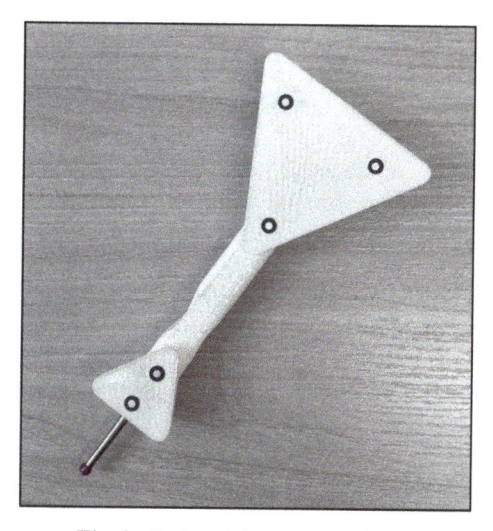

Fig. 2. Probe with passive markers.

Therefore, it was decided to choose active markers - diodes placed in conical holes on the surface of the probe (Fig. 3).

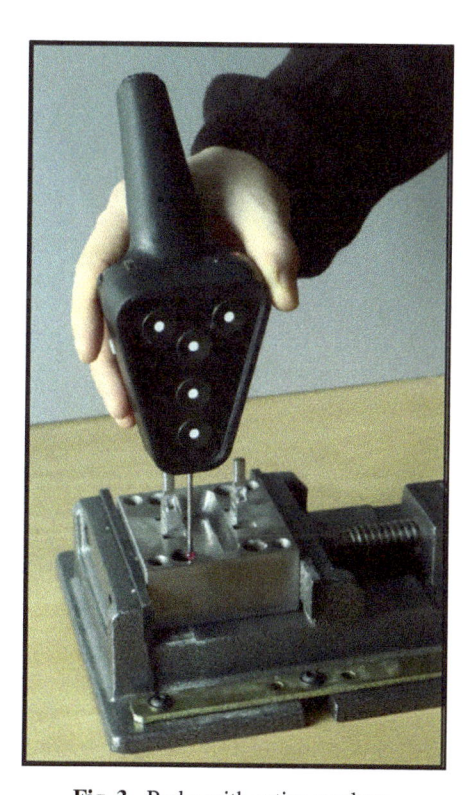

Fig. 3. Probe with active markers.

The next aspect considered was the shape of the probe. Various variants of the measuring probe were developed: due to their design and the arrangement of markers on the probe. Work was carried out on ergonomics, functionality and the appearance of the probe.

The next stage was to calibrate the probe. In the probe with active markers, the measurement of markers was performed with the Leitz PMM 12106 with increased MPE = 0.6 μm + 0.7 L μm/m with a contact probe head. After mounting the matting plate, a stand for measurements was made so that the plate would not bend. Then, the markers were measured from both sides of the probe to determine the centers of the markers (Fig. 4). These coordinates were created at the intersection of the axis of the inner cone with the matting plate.

Each measurement was performed using an accredited method of calibrating objects, the so-called multi-position method. This method consists of measuring an element in four positions: base, object rotated about the x axis, object rotated about the y axis, object rotated about the z axis. Probes with active markers were calibrated with an uncertainty of 1.1 μm.

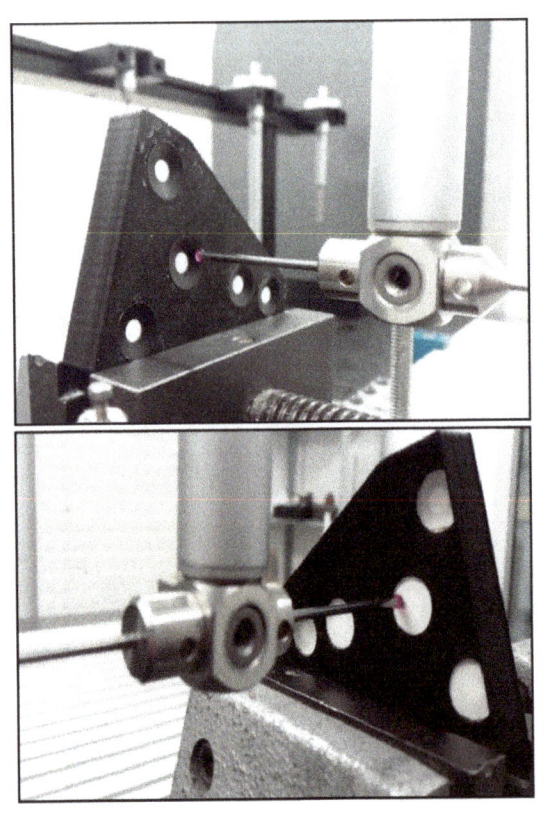

Fig. 4. Probe with active markers measured by WMP PMM.

An important element in the development of the probe is to define the algorithm of the x, y, z coordinates of the probe tip center. It must be calculated in relation to the centers of the markers obtained from CMM (Coordinate Measuring Machine) measurements. Measurements carried out on coordinate measuring machine was a key to improvement the accuracy of the measuring probe [10, 11]. For further research, it was necessary to develop a computer program allowing to determine the probe tip center based on the markers.

3 Validation of the Alignment of Coordinate System

In order to increase the accuracy of the measuring probe, a number of tests were carried out to check the algorithms used by the probe. The alignment of the coordinate system has been validated. The validation consisted in converting 4 points measured in the coordinate system (local system B) using the contact probe to the coordinate system of the scanner (global system A). The standard shown in Fig. 5 was used during the research. This standard consists of 3 balls and 3 cones.

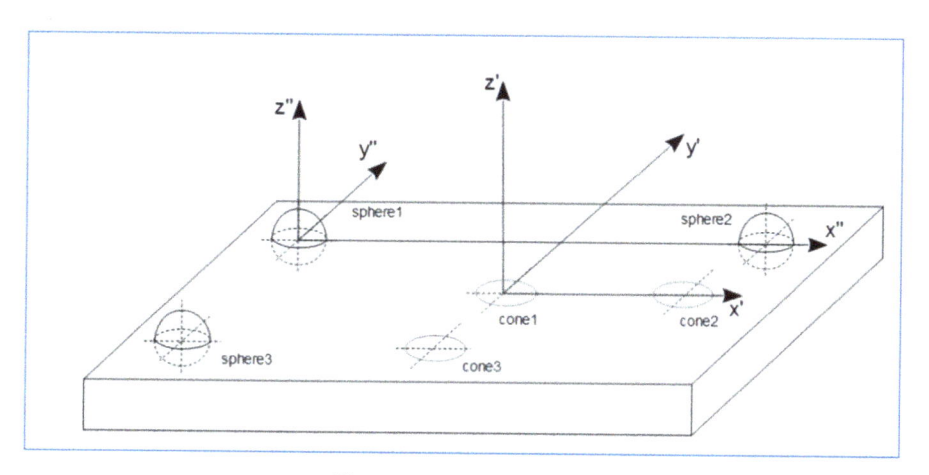

Fig. 5. Validation standard.

Two possible methods of solving the problem were validated:

- method 1 - in this method, using a contact probe, the centers of three spheres (Fig. 5.) and the points to be converted to the coordinate system of the scanner are measured. On the other hand, in the coordinate system of the scanner, the centers of three spheres should be determined, which enables the conversion of points from the contact system to the contactless system.
- method 2 - in this method, three cones (one point in each of the three cones) and the points to be converted to the coordinate system of the scanner are measured using the contact probe. On the other hand, in the coordinate system of the scanner, the centers of three spheres should be determined, which enables the conversion of points from the contact system to the contactless system.

120   R. Kupiec et al.

4 Validation of the Developed Program

VALIDATION METHOD 1.

Four points were measured using a contact probe (local system B) (Table 1):

Table 1. Input data.

	x [mm]	y [mm]	z [mm]
1	−114.08241	145.11110	481.95486
2	16.65738	135.72185	828.41166
3	216.00668	140.53416	719.08103
4	−3.15570	147.95254	417.38023

These points were entered into the reference software (PC-DMIS). They were then transformed according to method 1. The Table 2 shows the coordinates of the four points after conversion in the reference software (PC-DMIS). These results were taken as reference results (PC-DMIS software is certified metrology software).

Table 2. Reference results.

	x [mm]	y [mm]	z [mm]
1	110.58510	−118.43115	100.41587
2	189.05822	91.48571	−194.52522
3	395.10355	93.15632	−98.30144
4	225.75765	−119.41121	157.13459

The next step was to recalculate the input data (Table 1) in the validated software. The converted coordinates of the points obtained from the validated software are presented in Table 3. The differences between these results and the reference results are listed in the Table 4.

Table 3. Results from validated software.

	x [mm]	y [mm]	z [mm]
1	110.58509	−118.43115	100.41587
2	189.05823	91.48572	−194.52522
3	395.10356	93.15631	−98.30144
4	225.75765	−119.41122	157.13459

Table 4. Differences between the results of the validated program and the reference results (method 1).

	x [mm]	y [mm]	z [mm]
1	−0.00001	0.00000	0.00000
2	0.00001	0.00001	0.00000
3	0.00001	−0.00001	0.00000
4	0.00000	−0.00001	0.00000

The largest differences observed were ± 0.01 µm. Software errors can be considered as irrelevant.

VALIDATION METHOD 2.

The same input data presented in Table 1 was used to validate method 2. And the same reference data (Table 2). The next step was to recalculate the input data in the validated software. The converted coordinates of the points obtained from the validated software are presented in Table 5. Table 6 shows the differences between the results from Tables 2 and 5.

Table 5. Results from validated software.

	x [mm]	y [mm]	z [mm]
1	110.04645	−118.47225	100.11343
2	190.78408	91.61205	−194.09633
3	396.07815	93.24531	−96.27933
4	224.77652	−119.47480	157.72157

Table 6. Differences between the results of the validated program and the reference results (method 2).

	x [mm]	y [mm]	z [mm]
1	−0.53865	−0.04110	−0.30244
2	1.72586	0.12634	0.42889
3	0.97460	0.08899	2.02211
4	−0.98113	−0.06359	0.58698

The biggest difference that was observed between the results obtained from the validated program and the reference results in method 2 was 2.022 mm. The errors of the second method are much bigger than that of the first method.

5 System Calibration Procedure

To check the accuracy of the measurement, an accredited calibration procedure of the optical system was used, based on VDI/VDE 2634. This section presents the accuracy of the contact head. According to VDI/VDE, parameters are measured: probing error (form) PF, probing error (size) PS, sphere spacing error SD and flatness measurement error F [12, 13].

The probing error describes a characteristic error in 3D optical measuring systems based on surface scanning in a small measuring range. It is the distance between the center of the sphere determined using the Gaussian criterion (the method of least squares) [14].

As a standard, spheres made of ceramics, steel or other materials adequately diffusing light with a diameter of:

$$d = (0, 1 \ldots 0, 02) \bullet L_0 \tag{1}$$

where:

L_0 - diagonal of the smallest rectangular parallelepiped encompassing the measurement space. The procedure consists in measuring the ball in at least 10 settings throughout the measurement space (Fig. 6).

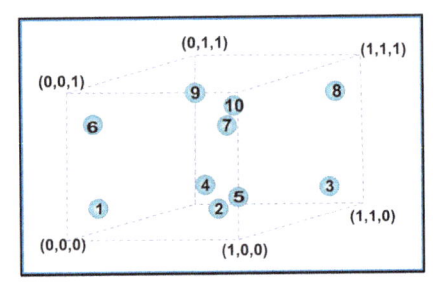

Fig. 6. Distribution of the test ball throughout the measurement space of the system during the measurement.

Another qualitative parameter, the sphere spacing error, is used to verify the ability of the system to reconstruct the length. It is the difference between the measured value and the calibrated distance between the balls:

$$\Delta l = lm - lk \tag{2}$$

where:
Δl - error of the distance between the centers of the spheres,
lm - measured length value,
lk - calibrated length value.

To test the error of the distance between the centers of the spheres, a ball-bar standard witch two spheres made of steel, ceramics or other appropriate materials can be used. In order to investigate the error of indication along the length, the standard is measured in six settings (Fig. 7).

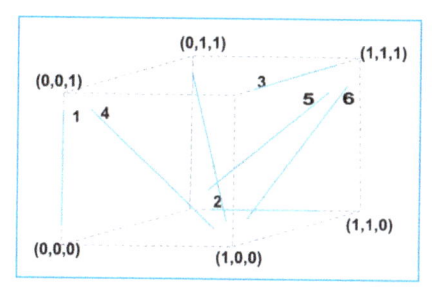

Fig. 7. The arrangement for determination of the sphere spacing error.

Flatness measurement error is defined as the range of distances of the points measured from the plane constructed according to the least squares method by the best fit. Standards in the form of cuboids made of steel, ceramics, aluminum or other material with low reflectivity are used here, the width of which cannot be less than 50 mm, and length than $0.5 \bullet L_0$. In order to determine the flatness measurement error, a measurement should be made in min. six settings (Fig. 8).

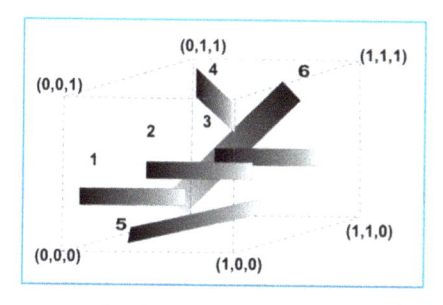

Fig. 8. The arrangement for determination of the flatness measurement error.

6 Results of Calibration

A reference ball was used to determine the error of the probing error (form) PF and the probing error (size) PS. During the measurement, this sphere was positioned so as to cover the entire measuring space. Each time, 25 points were collected evenly distributed on the sphere. As can be seen from the values presented in Table 1, MPE (PF) is 0.031 mm, while MPE (PS) is 0.026 mm. MPE values take into account the measurement uncertainty of the standard during its calibration (Table 7).

Table 7. The results of the determination of the probing error. Nominal value of the reference ball diameter is 24.98093 mm.

Position	PF [mm]	U(PF) [mm]	Measured diameter [mm]	PS [mm]	U(PS) [mm]
1	0.0241	0.0014	24.9717	0.0092	0.0006
2	0.0173	0.0014	24.9778	0.0031	0.0006
3	0.0183	0.0014	24.9608	0.0201	0.0006
4	0.0292	0.0014	24.9612	0.0197	0.0006
5	0.0021	0.0014	24.9590	0.0219	0.0006
6	0.0119	0.0014	24.9554	0.0255	0.0006
7	0.0277	0.0014	24.9622	0.0187	0.0006
8	0.0176	0.0014	24.9608	0.0201	0.0006
9	0.0211	0.0014	24.9747	0.0062	0.0006
10	0.0282	0.0014	24.9717	0.0092	0.0006

Max(PF) 0.0292 mm Max(PS) 0.0255 mm.
MPE(PF) 0.0310 mm MPE(PS) 0.0260 mm.

Another parameter checked was the sphere spacing error SD. Setting the standard during the measurement was consistent with Fig. 6. For each of the 7 measurements, the parameter was calculated as the absolute value of the difference between the measured value Lz and the nominal value Ln, SD = |$Lz - Ln$|. The MPE (SD) value was determined at the level of 0.0290 mm (Table 8).

Table 8. Measurement results of the length standard.

Position	L_z [mm]	L_n [mm]	SD [mm]	U [mm]
X	269.3813	269.4036	0.0223	0.0007
Y	269.3893	269.4036	0.0143	0.0007
Z1	269.3955	269.4036	0.0081	0.0007
D1	269.4247	269.4036	−0.0211	0.0007
D2	269.4234	269.4036	−0.0198	0.0007
D3	269.4254	269.4036	−0.0218	0.0007
D4	269.4315	269.4036	−0.0279	0.0007

Max(SD) 0.0279 mm.
MPE(SD) 0.0290 mm.

The last of the determined parameters was the flatness error, where the standard was set in the measurement space as shown in Fig. 7. The maximum flatness error was determined at the level of MPE (F) = 0.034 mm (Table 9).

Table 9. Flatness error results.

Position	F [mm]	U [mm]
1	0.0138	0.0026
2	0.0232	0.0026
3	0.0312	0.0026
4	0.0282	0.0026
5	0.0186	0.0026
6	0.0277	0.0026

Max(F) 0.0312 mm.
MPE(F) 0.0340 mm.

All the parameters of the tested system have been determined according to VDI/VDE 2634. The parameters such as maximum permissible errors (MPE) was determined.

7 Conclusions

Optical measurements are very important in industry today. They are very profitable from an economic point of view. We can measure many features at once in a short period of time. However, 3D scanners are not able to measure all dimensions. Sometimes the measurements require collecting points in places not visible by the 3D scanner. The contact measuring probe is more suitable for such applications.

Many problems have been solved during the research. Passive markers are easier to apply than active markers. On the other hand, active markers allow you to achieve more accurate results.

The correct determination of the marker centers allowed to improve the accuracy of the measuring probe. This was made possible by measuring the marker centers on coordinate measuring machine.

The measuring system consisting of the 3D scanner and the handheld measuring probe was successfully tested according to VDI/VDE 2634 [14]. The system parameters were determined in accordance with metrological requirements and the measurement uncertainty was taking into account.

The software module responsible for the alignment of coordinate systems was validated. The software enables alignment by two methods. The validation of method one showed that the software had negligibly small errors. The validation of the second method showed that the errors are much larger than for the first method, so it is recommended to use method 1 as it gives less errors. The biggest limitation of using handheld probe is the reduced measuring space in relation to the space of the entire system. If the markers on the probe (even one) are invisible to the system, the algorithm is unable to calculate where the measuring tip is located.

Further work will involve the preparation of a correction matrix for the described probe, using reference measuring machines.

Acknowledgments. This work was supported by Ministry of Economic Development in Poland grant number: POIR.01.01.01-00-0376/15, co-financed by Smart Growth Operational Programme in the years 2014-2020.

References

1. Owczarek, D., Ostrowska, K., Sładek, J.: Examination of optical coordinate measurement systems in the conditions of their operation. J. Mach. Constr. Maintenance Probl. Eksploatacji **4**, 7–19 (2017)
2. Owczarek, D., Ostrowska, K., Harmatys, K., Sładek, J.: Estimation of measurement uncertainty with the use of uncertainty database calculated for optical coordinate measurements of basic geometry elements. Adv. Sci. Technol. Res. J. **9**(27), 112–117 (2015). https://doi.org/10.12913/22998624/59092
3. Juras, B., Szewczyk, D.: Dokładność pomiarów realizowanych skanerem optycznym, Postępy Nauki i Techniki, Politechnika Lubelska 7 (2011)
4. Urbas, U., Zorko, D., Črne, B., Tavčar, J., Vukašinović, N.: A method for enhanced polymer spur gear inspection based on 3D optical metrology. Measurement **169** (2021)
5. Helle, R.H., Lemu, H.G.: A case study on use of 3D scanning for reverse engineering and quality control. Mater. Today Proc. **45**(Part 6), 5255–5262 (2021)
6. Sładek, J.: Dokładność pomiarów współrzędnościowych, Wydawnictwo Politechniki Krakowskiej, Kraków (2011)
7. Sitnik, R.: Odwzorowanie kształtu obiektów trójwymiarowych z wykorzystaniem oświetlenia strukturalnego, Oficyna Wydawnicza Politechniki Warszawskiej (2010)
8. Sitnik, R.: New method of structure light measurement system calibration based on adaptive and effective evaluation of 3D-phase distribution. Proc. SPIE, Opt. Meas. Syst. Ind. Inspection IV, **5856**, 109–117 (2005)
9. Song, M., Chen, F., Brown, G.M.: Overview of 3-D shape measurement using optical methods. Opt. Eng. **39**, 10–22 (2000)
10. Kupiec, M.: Optyczno-stykowa metoda pomiarów współrzędnościowych, Ph.D. thesis, Politechnika Krakowska (2007)
11. Bernal, C., de Agustina, B., Marín, M.M., Camacho, A.M.: Performance evaluation of optical scanner based on blue LED structured light. Procedia Eng. **63**, 591–598. The Manufacturing Engineering Society International Conference, MESIC (2013)
12. Bonin, R., Khameneifar, F., Mayer, J.R.R.: Evaluation of the metrological performance of a handheld 3D laser scanner using a pseudo-3D ball-lattice artifact. Sensors (2021)
13. Ostrowska, K., Szewczyk, D., Sładek, J.: Optical systems calibration according to ISO standards and requirements of VDI/VDE. Wzorcowanie systemów optycznych zgodnie z normami ISO i zaleceniami VDI/VDE. Technical Transactions. Mechanics (2012)
14. VDI-VDE-2634-Systeme mit flaechenhafter Antastung (2013)

Comparison of Measurements Realized on Computed Tomograph and Optical Scanners for Elements Manufactured by Wire Arc Additive Manufacturing

Michał Wieczorowski[1], I. P. Yago[2], Pereira Domínguez Alejandro[2], Bartosz Gapiński[1(✉)], Grzegorz Budzik[3], and Magdalena Diering[1,2,3]

[1] Poznan University of Technology, M. Skłodowskiej 5 sq., 60965 Poznan, Poland
{michal.wieczorowski,bartosz.gapinski,
magdalena.diering}@put.poznan.pl
[2] University of Vigo, Circunvalación ao Campus Universitario, 36310 Vigo, Pontevedra, Spain
apereira@uvigo.es
[3] Rzeszów University of Technology, al. Powstańców Warszawy 12, 35959 Rzeszów, Poland
gbudzik@prz.edu.pl

Abstract. Additive techniques become more and more common in manufacturing processes. Among metallic materials an interesting technique for depositing metallic layers is the use of Wire Arc Additive Manufacturing process, where 3D metallic structures are created using welding technologies, i.e. Gas Metal Arc Welding. In the paper an analysis of measurement devices for surfaces after that kind of manufacturing was presented. A computer tomograph as well as two types of scanners were used, respectively with a high and low resolution. For dimensional measurements the results showed that a CT is a good option, enabling to properly represent the real work piece. The results obtained with a high resolution scanner were usually pretty close, except for few cases. On the other hand, a low resolution scanner due to a large distance between points was not able to show good dimensions. Pores in structures were also inspected. The biggest problems with pores occurred where path of a robotic arm was the most complicated.

Keywords: Wire Arc Additive Manufacturing · Scanner · Computed tomography

1 Introduction

Additive techniques are gaining popularity in manufacturing processes [1–3]. Both techniques using plastics [4] and metal materials [5] are used. The multitude of techniques associated with manufacturing include methods of Selective Laser Sintering (SLS), Digital Light Processing (DLP), Fused Filament Fabrication (FFF), Direct Metal Laser Sintering (DMLS), VAT Photopolymerization (VPP), Material Jetting Additive (MJT), Binder Jetting (BJT), Powder Bed Fusion (PBF), Material Extrusion (MEX) or Directed

Energy Deposition (DED). Among metallic materials, powder sintering [6], laser [7] or electron beam forming [8] and deposition [9] are the dominant methods. In the latter group of methods, an interesting technique for depositing metallic layers is the use of Wire Arc Additive Manufacturing (WAAM) process [10], where 3D metallic structures are created using welding technologies [11], i.e. Gas Metal Arc Welding (GMAW) [12]. Generally speaking, Additive Layer Manufacturing, including Wire and Arc Additive Layer Manufacture (WAALM) became an interesting and important technology which is more and more often used in industrial processes. It can be used to manufacture work pieces out of many different materials, answering a demand for sustainable, low cost and environmentally friendly manufacturing processes with very high geometric flexibility. Deposition of materials with Wire and Arc Additive Layer Manufacture comprise of e.g. steel [13], Ni alloys [14], and Ti alloys [15]. The process can be run with various welding parameters [16] and - as a usual modern welding operation - is usually performed with a robot [17].

In order to demonstrate the methodology and measurement capabilities (digitalization) of the structures described above, various macro-scale measuring devices were used. The aim of the study was to determine the possibility of mapping thin walls and irregularities occurring on the surface, as well as pores in the material. This type of research is important from the point of view of practical applications of measurement systems in additive techniques and the associated need to determine the appropriate accuracy parameters.

2 Measurement Techniques in Additive Manufacturing

Additive techniques do not always allow the use of classical measurement methods [18]. Some features obtained as a result of them also require a special mathematical approach [19]. Measurements of structures obtained by these techniques on a macro scale more justify the use of measurement methods based on electromagnetic radiation than contact techniques. Therefore, in this paper, optical scanners of different resolution and a computer tomograph were used for measurements.

Coordinate measuring machines in macro scale have gained popularity over a number of years and have become indispensable in various industries. But coordinate measuring technology has gone much further in recent years. The variety of measurement tasks and complex shapes of workpieces have led to the development of optical measuring tools that allow to solve problems that are difficult to realize with contact methods. These include 3D scanners. Their great advantage is the speed of acquiring an enormous amount of information on the product being measured and the clarity of the presentation of the results. They are based on the principle of measuring the light reflected or scattered from the surface of the measured object and triangulation. We divide them basically into structural light scanners and laser ones. Structured light scanners [20, 21] are based on the projection of structured light, i.e. their projectors project patterns of e.g. Grey codes, phase shifted bars or other code forms (free shapes, circles, etc.) onto the measured surface. They are usually mounted on tripods. They are quite common in many different applications, including not only mechanical engineering [22], but also biomedicine [23], surface defects [24], concrete [25] or casting cores [26]. In many applications, structured light scanners are being increasingly replaced by newer solutions, namely laser scanners [27, 28]. Initially, solutions based on laser action were just laser scanning heads [29, 30] or laser triangulation heads, that were placed, for example, on CMMs or measuring arms. Currently, they function as stand-alone laser scanners [31] operating in manual or automatic mode and are one of the most common and effective data acquisition methods in a wide variety of applications [32], including also analysis of wire arc hybrid manufacturing process [33] or even very large work pieces [34]. The working principle of laser scanning heads is based on emitting a point, a line or a set of laser lines on the surface of the scanned object. These elements are observed by the CCD camera giving coordinates of measurement points. They avoid problems associated with vibration and scanning from the so-called "free hand" and are less sensitive to light reflections. A schematic picture of example laser scanners is shown in Fig. 1 [35].

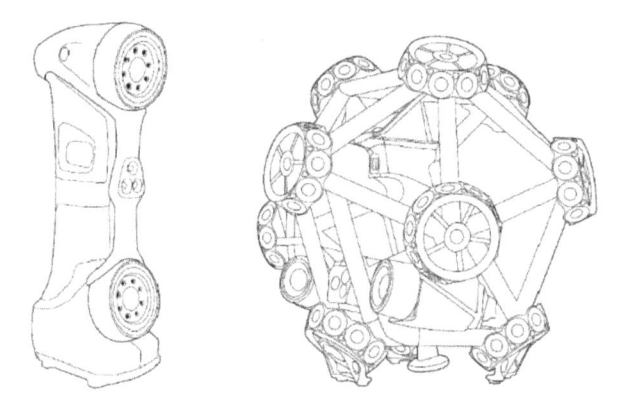

Fig. 1. Schematic image of laser scanners.

Electromagnetic radiation can be used in metrology not only in the visible range. Computed tomography [36], which is based on X-rays, is increasingly used not only to analyze defects but also geometrical features. Technical tomographs, allow imaging with a resolution even below a micrometer [37]. The classical system consists of a measured object placed on a rotating table and the lamp and detector that are stationary or perform linear motion. The radiation beam is attenuated as it passes through the object, and this process depends on the thickness of the absorbing medium and the absorption coefficient. During the measurement, a number (usually hundreds or thousands) of 2D X-ray images are taken for different angular positions of the lamp-detector system relative to the measured object [38]. As a result of reconstruction from 2D shots, a spatial image is obtained. Technical tomographs are used for the analysis of very different elements, including objects made of plastic [39], foam [40], and metal, also those made with additive techniques [41]. In this case, the analytical results are often better than those obtained by metallography [42]. Also, the accuracy parameters [43] and the results of comparisons with other measurement methods used in coordinate technology [44] make this application future of computed tomography look very positive.

3 Materials and Measures

A robotic additive manufacturing system using GMAW was used in this study. The test samples were made out of S235 JR steel using AWS ER70S-6 wire material and 15% CO_2 and 85% argon gas composition. The torch provided a flux of gas during the welding process and also during a brief period, before arc establishment and after the end of it. It ensures that the material is protected from oxidation every time. The samples were made with 40 A intensity, 5 V, 310 mm/min feed rate and 1,7 m/min wire speed.

There were three different shapes of samples, prepared to show various opportunities of GMAW. Their nominal forms are presented on Fig. 2.

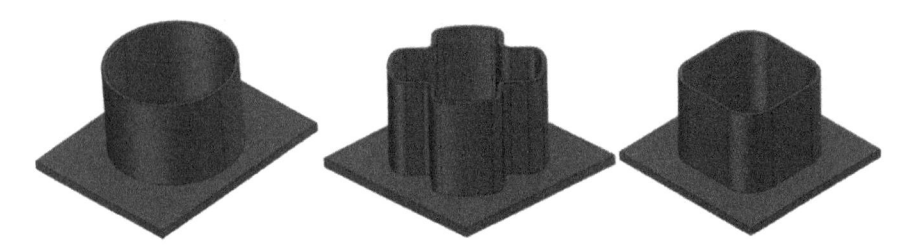

Fig. 2. Nominal shapes of samples.

For each type of shape (cylinder, cross and rectangle) three specimens were manufactured. The examples are presented on Fig. 3.

Fig. 3. Images of samples with cylindrical, cross and rectangle shape.

Cylinders were prepared to a nominal inner diameter of 22 mm and an outer diameter of 33 mm, assuming maximum inscribed element for inner diameters and minimum circumscribed element for outer diameter. Crosses had an inner diameter of 18 mm and an outer diameter of 33 mm, evaluated the same way. For rectangles, X and Y directions were determined, for which the nominal dimensions were: outer dimension X - 28 mm, inner dimension X - 21 mm, outer dimension Y - 26 mm, inner dimension Y - 19 mm.

To investigate possibilities of digitization three different measurement devices were used; two optical structural light scanners and a computer tomograph. The scanners were selected with different resolutions, to verify an influence of resolution on image and data quality. The low resolution scanner (LRS) had a resolution of 0,5 mm, while the high resolution scanner (HRS) enabled to collect data with 0,115 mm measuring point distance. These values imply the minimum size of 3D structures that may be identified by each of scanners. For computed tomography (CT) measurements, a 320 W directional open-type lamp with a 200 μm pixel size detector was used. Measurements were performed at 230 kV and 200 μA current, and 1000 images were taken per rotation. Data were collected with a voxel size of 50,375 μm.

4 Results and Discussion

The first study examined the differences in shape representation and dimensions for each type of work piece. All the samples were aligned to their nominal shapes using a best fit method. The alignments were shown on Fig. 4.

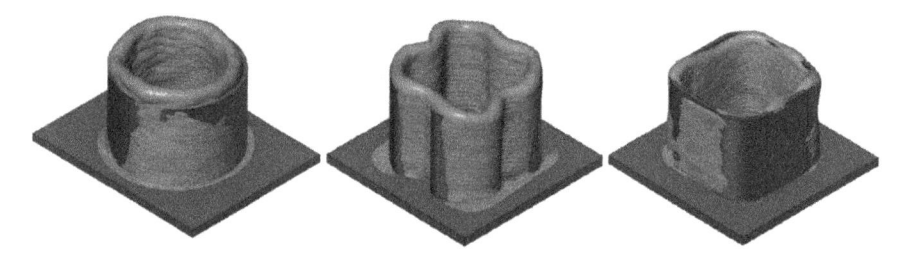

Fig. 4. Examples of alignment for different geometries: cylinder, cross and rectangle.

The results obtained for the cylinders are presented in Tables 1 and 2.

Table 1. Comparison of inner diameter (in mm) for cylinders (nominal 22 mm).

	Computed Tomograph CT	High Res Scanner HRS	Low Res Scanner LRS
Cylinder 1	21,607	21,592	21,743
Cylinder 2	23,823	23,855	23,772
Cylinder 3	22,521	22,538	23,089

Table 2. Comparison of outer diameter (in mm) for cylinders (nominal 33 mm).

	Computed Tomograph CT	High Res Scanner HRS	Low Res Scanner LRS
Cylinder 1	33,089	33,167	32,808
Cylinder 2	32,805	32,831	33,642
Cylinder 3	33,875	33,830	33,296

The images of work pieces are presented on Fig. 5.

Fig. 5. Examples of diameter measurement for cylinders, from left to right: computed tomograph, high resolution scanner, low resolution scanner.

As can be seen from the presented data, the differences in dimensions between HRS and CT except for one case do not exceed 30 μm. However, for LRS the deviations are much larger and exceed even 0.5 mm. The images of cylinders retrieved by measuring devices show similar shapes for CT and HRS, while LRS had problems with identification of upper regions.

The same procedure was applied to cross shaped elements. The obtained results are presented in Tables 3 and 4.

Table 3. Comparison of inner diameter (in mm) for crosses (nominal 18 mm).

	Computed Tomograph CT	High Res Scanner HRS	Low Res Scanner LRS
Cross 1	17,706	17,676	18,396
Cross 2	18,268	18,296	18,793
Cross 3	18,272	18,255	18,946

Table 4. Comparison of outer diameter (in mm) for crosses (nominal 33 mm).

	Computed Tomograph CT	High Res Scanner HRS	Low Res Scanner LRS
Cross 1	33,088	33,120	32,506
Cross 2	33,184	33,191	32,478
Cross 3	33,061	33,087	32,078

The images of work pieces are presented on Fig. 6. For these surfaces, the dimensional difference between HRS and CT is similar to that for cylinders and ranges from 7 to 32 μm. For LRS the deviations are again much larger and reach up to 1 mm. The images of surfaces obtained by measuring devices similarly as before show similar shapes for CT and HRS, also here LRS had problems with identification of upper fragments, although smaller than in case of cylindrical surfaces. This is due to the fact that the surfaces are wider in places where the robot changes direction of its movement, making the LRS measurement task easier.

Fig. 6. Examples of diameter measurement for crosses, from left to right: computed tomograph, high resolution scanner, low resolution scanner.

A measurement procedure for rectangles included measurements of internal and external dimensions in two mutually perpendicular directions, perpendicular also to the sides of the figures. The obtained results are presented in Tables 5 and 6.

Table 5. Comparison of inner dimensions (in mm) for crosses (nominal X = 21 mm; Y = 19 mm).

	Computed Tomograph CT		High Res Scanner HRS		Low Res Scanner LRS	
	X	Y	X	Y	X	Y
Rectangle 1	21,099	18,840	21,401	18,113	22,171	20,004
Rectangle 2	20,932	18,784	20,698	19,188	21,519	20,077
Rectangle 3	20,858	18,483	21,105	19,216	21,603	20,020

Table 6. Comparison of outer dimensions (in mm) for crosses (nominal X = 28 mm; Y = 26 mm).

	Computed Tomograph CT		High Res Scanner HRS		Low Res Scanner LRS	
	X	Y	X	X	Y	X
Rectangle 1	27,569	26,116	27,529	26,285	26,841	25,557
Rectangle 2	28,111	26,773	27,672	26,532	26,958	25,345
Rectangle 3	28,058	26,444	27,538	26,173	26,925	25,409

The images of work pieces are presented on Fig. 7.

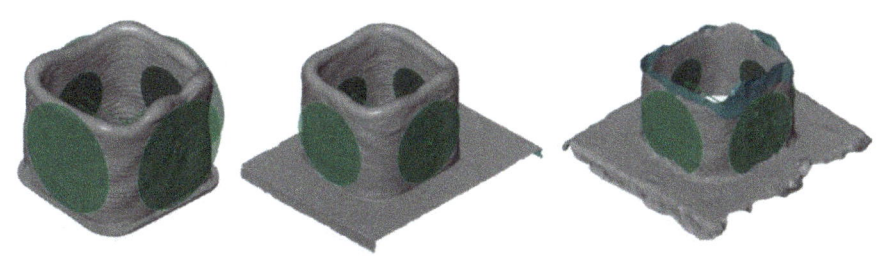

Fig. 7. Examples of diameter measurement for rectangles, from left to right: computed tomograph, high resolution scanner, low resolution scanner.

For these surfaces, the inner and outer dimension were analyzed in two directions. For the inner dimension, the differences between HRS and CT were large, up to 0.7 mm in the Y axis. The LRS on the other hand differed from the CT even more, up to about 1.5 mm. These differences show that for both scanners, the correct representation of the inner surface of a relatively small hole was a difficulty, which is natural when triangulation of this type of surface is concerned. The differences in the outer surfaces measured by CT and HRS were smaller and in most cases did not exceed 0.2 mm, which however is still significantly larger than the corresponding differences for the cylindrical and cross shaped surfaces. The reason for this phenomenon can be seen in the surface images, because HRS much less reliably reproduced the irregularities of the outer surface, i.e., phenomena occurring at the meso and micro scale. LRS also in this case showed large differences from the dimensions obtained with CT, with values reaching almost 1.5 mm. Also on these objects the LRS had great problems in identifying the top fragments.

From the images, it can be seen, that the thickness is growing when the layers are superposing. Due to that, there is the thinnest part in every sample located close to the plate. The values of thickness were calculated for each sample, as maximum, minimum and average. For cylinders, the results are presented in Table 7.

Table 7. Comparison of thickness values (in mm) for a cylinder.

	Computed Tomograph CT	High Res Scanner HRS	Low Res Scanner LRS
Maximum	3,461	3,250	3,413
Minimum	1,317	1,835	2,061
Average	2,663	2,715	2,625

The images of work pieces are presented on Fig. 8.

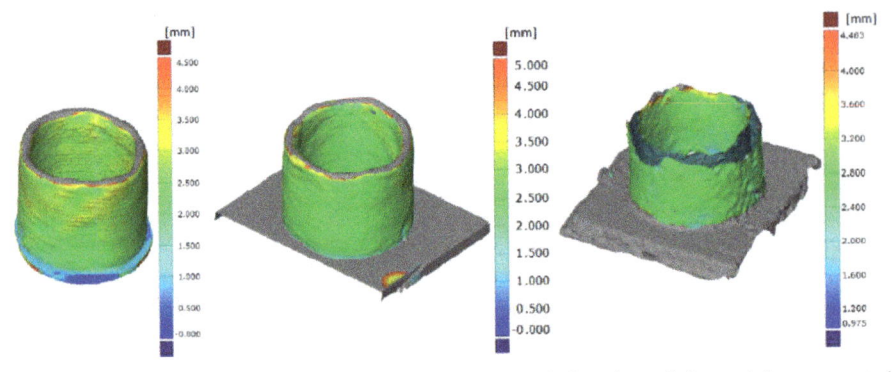

Fig. 8. Examples of thickness measurement for the cylinder, from left to right: computed tomograph, high resolution scanner, low resolution scanner.

The obtained differences show relatively good convergence for all three devices for maximum and average thickness. Bigger differences can be seen in the minimum thickness, and the larger values for HRS and LRS than for CT show that the scanners cannot quite good represent some of the valleys on the surfaces, tending to filter them out (LRS obviously filters out more). The surface images similarly show that the LRS resolution was not sufficient to represent the wall thickness of the cylinder.

The same procedure was also applied to cross shaped elements. The obtained results are presented in Table 8.

Table 8. Comparison of thickness values (in mm) for a cross shaped element.

	Computed Tomograph CT	High Res Scanner HRS	Low Res Scanner LRS
Maximum	3,592	3,553	3,003
Minimum	1,149	1,790	2,036
Average	2,881	2,895	2,490

The images of the cross shaped work piece are presented on Fig. 9.

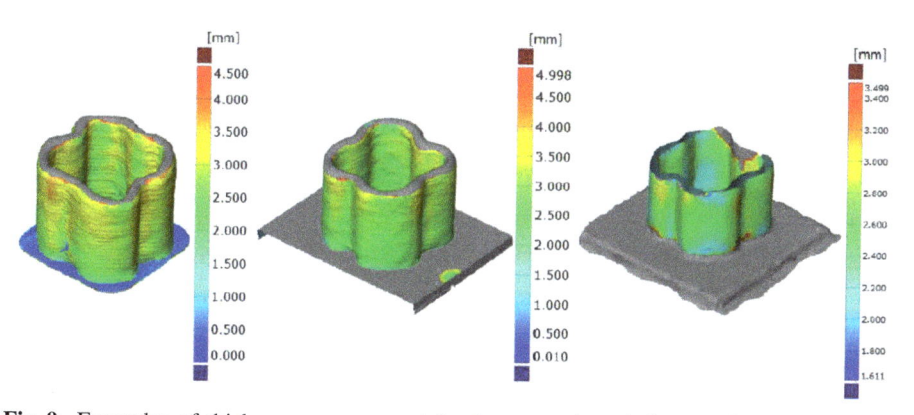

Fig. 9. Examples of thickness measurement for the cross shaped element, from left to right: computed tomograph, high resolution scanner, low resolution scanner.

The differences obtained show a relatively good convergence for CT and HRS for maximum and average thickness, in this case LRS showed smaller values, what - as previously - is related to the difficulty of mapping a complex surface. Big differences can again be seen in the minimum thickness, and here too larger values were obtained for HRS and LRS than for CT. On cross shaped surfaces there are more valleys and pitches, making it even more difficult for the scanners to image correctly, causing a filtering effect (again the LRS filters more). In the surface images, also in this case, the problems of LRS with the resolution and - consequently - the representation of the wall thickness of the element are clearly noticeable.

An analogous procedure was used to measure the wall thickness of elements made in rectangular shape. The results obtained for this element are presented in Table 9.

Table 9. Comparison of thickness values (in mm) for a rectangle.

	Computed Tomograph CT	High Res Scanner HRS	Low Res Scanner LRS
Maximum	3,685	3,345	3,097
Minimum	0,865	1,810	1,815
Average	2,813	2,737	2,329

The images of the cross shaped work piece are presented on Fig. 10.

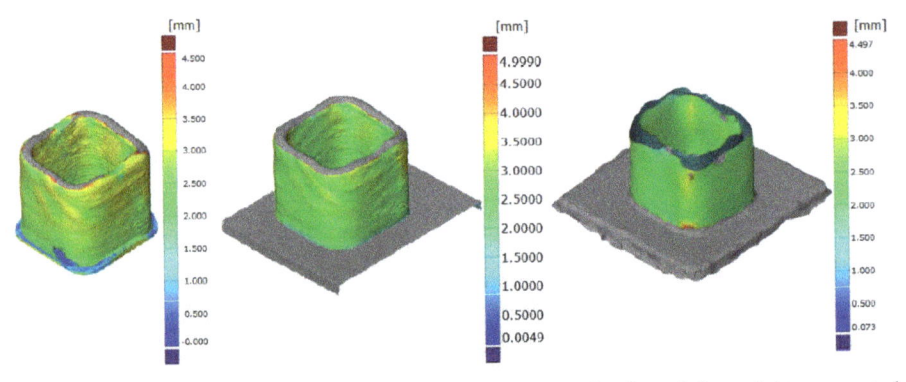

Fig. 10. Examples of thickness measurement for the rectangle, from left to right: computed tomograph, high resolution scanner, low resolution scanner.

The differences obtained show good convergence for CT and HRS for the average thickness, a little worse for the maximum thickness. The image obtained from HRS shows a very smoothed outer surface compared to reality. The LRS again showed smaller values related to mapping difficulty, although the differences are less significant than for the previous surfaces. The big differences are again in the minimum thickness, with larger values for the scanners. Interestingly, both showed very similar values, which may indicate similar problems in identifying the interior of the object. As in all cases there are also LRS problems with mapping the wall thickness of the part.

A crucial feature that is related with an additive process of production, is the porosity. For that reason, an analysis of the entire sample using a CT was performed. What is clear, this analysis is not possible with optical scanners. Table 10 shows the volume of voids that the was found during the scanning process. In the table, vacuum volume (mm^3) means the summation of voids volume in the whole specimen, vacuum volume (%) represents the percentage of the summation of voids volume in relation to the total sample volume, and maximum pore diameter shows maximum diameter of the biggest internal void.

Table 10. Porosity of samples.

	Vacuum volume (mm^3)	Vacuum volume (%)	Maximum pore diameter (mm)
Cylinder 1	No voids	No voids	No voids
Cylinder 2	No voids	No voids	No voids
Cylinder 3	No voids	No voids	No voids
Cross 1	0,001	0,000	0,194
Cross 2	5,176	0,093	2,120
Cross 3	No voids	No voids	No voids
Rectangle 1	No voids	No voids	No voids
Rectangle 2	0,018	0,001	0,460
Rectangle 3	No voids	No voids	No voids

The X-ray images of the samples with pores were presented on Fig. 11.

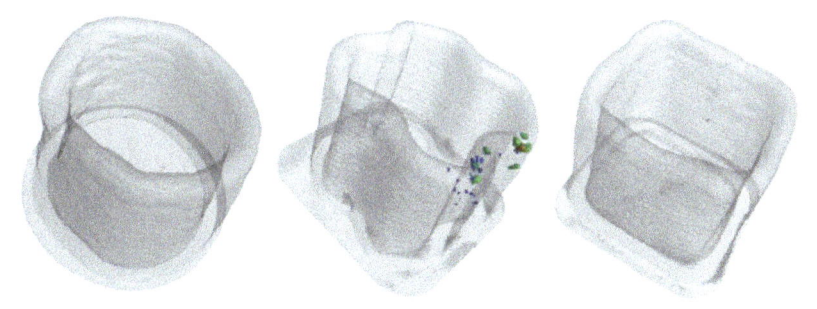

Fig. 11. X-ray images of pores for different geometries: cylinder, cross and rectangle.

The analysis of the occurrence and size of the pores showed that the process was mostly correct and therefore the number and size of pores is not large. For cylindrical surfaces, the easiest for the depositing robot, no pores were found at all. It should be noted, however, that the power of CT lamp required to overexpose the steel caused the resolution to deteriorate, making the detection of very small pores impossible. The largest number of pores was observed for cross shaped surfaces, the most difficult from the point of view of the kinematics of the metal depositing system.

5 Conclusions

The presented measurement results have shown the possibilities of applying measurement techniques using electromagnetic waves to the analysis of structures made with the GMAW method. These structures - as shown by the samples - allow relatively free shape forming, especially if the surfacing head is guided by a robotic arm.

The CT scanner made it possible to reflect the shape of the measured parts and performed the measurement task very well. X-rays moving through the measured object made it possible to find even narrow pockets, naturally occurring after the application of layers in the GMAW technology. It showed also the pores occurring in the structures and in this respect CT is the only non-destructive measuring method. Optical scanners were able to show external structures to varying degrees. The high-resolution scanner reproduced the shape of the objects relatively faithfully. However, it had problems with mapping internal surfaces and in some cases external ones at meso and macro scales, which unfortunately also had some effect on the macro scale. The low resolution scanner, on the other hand, proved to be essentially useless for this application, the dimensions obtained with it were only a very rough estimate of reality. Thin walls made representation of the height of the analyzed elements very difficult, as the measurement device had a problem with detecting the upper part of the walls. This shows how the resolution of the scanner affects the reliability of the reproduced shapes.

Another issue that can be observed is generally rather low repeatability of the welding work blank. In GMAW it is an effect of the difficulty of controlling the arc forming position. This is why the element is usually machined afterwards in the and a hybrid manufacturing process is applied.

Acknowledgments. This research was funded by the Polish Ministry of Higher Education grants No. 0614/SBAD/1547.

References

1. Shah, J., Snider, B., Clarke, T., Kozutsky, S., Lacki, M., Hosseini, A.: Large-scale 3D printers for additive manufacturing: design considerations and challenges. Int. J. Adv. Manuf. Technol. **104**(9–12), 3679–3693 (2019). https://doi.org/10.1007/s00170-019-04074-6
2. Sathies, T., Senthil, P., Anoop, M.S.: A review on advancements in applications of fused deposition modelling process. Rapid Prototyping J. **26**(4), 669–687 (2020). https://doi.org/10.1108/RPJ-08-2018-0199
3. Prater, T., Werkheiser, N., Ledbetter, F., Timucin, D., Wheeler, K., Snyder, M.: 3D Printing in zero G technology demonstration mission: complete experimental results and summary of related material modeling efforts. Int. J. Adv. Manuf. Technol. **101**(1–4), 391–417 (2018). https://doi.org/10.1007/s00170-018-2827-7
4. Rokicki, P., et al.: Rapid prototyping in manufacturing of core models of aircraft engine blades. Aircr. Eng. Aerosp. Technol. **86**(4), 323–327 (2014). https://doi.org/10.1108/AEAT-10-2012-0192
5. Pisula, J.M., Budzik, G., Przeszłowski, Ł.: An analysis of the surface geometric structure and geometric accuracy of cylindrical gear teeth manufactured with the direct metal laser sintering (DMLS) method. Strojniški vestnik – J. Mech. Eng. **65**(2), 78–86 (2018). https://doi.org/10.5545/sv-jme.2018.5614
6. Śliwa, R.E., Bernaczek, J., Budzik, G.: The application of direct metal laser sintering (DMLS) of titanium alloy powder in fabricating components of aircraft structures. Key Eng. Mater. **687**, 199–205 (2016). https://doi.org/10.4028/www.scientific.net/KEM.687.199
7. Santos, E.C., Shiomi, M., Osakada, K., Laoui, T.: Rapid manufacturing of metal components by laser forming. Int. J. Mach. Tools Manuf. **46**, 1459–1468 (2006)

8. Wanjara, P., Brochu, M., Jahazi, M.: Electron beam freeforming of stainless steel using solid wire feed. Mater. Des. **28**, 2278–2286 (2007)
9. Ribeiro, F.: 3D printing with metals. Comput. Control Eng. J. **9**(1), 31–38 (1998). https://doi.org/10.1049/cce:19980108
10. Xia, C., et al.: A review on wire arc additive manufacturing: monitoring, control and a framework of automated system. J. Manuf. Syst. **57**, 31–45 (2020). https://doi.org/10.1016/j.jmsy.2020.08.008
11. González, J., Rodríguez, I., Prado-Cerqueira, J-L., Diéguez, J.L., Pereira, A.: Additive manufacturing with GMAW welding and CMT technology. Procedia Manuf. **13**, 840–847 (2017). https://doi.org/10.1016/j.promfg.2017.09.189
12. Ding, J., et al.: Thermo-mechanical analysis of Wire and Arc Additive Layer Manufacturing process on large multi-layer parts. Comput. Mater. Sci. **50**(12), 3315–3322 (2011). https://doi.org/10.1016/j.commatsci.2011.06.023
13. Spencer, J.D., Dickens, P.M., Wykes, C.M.: Rapid prototyping of metal parts by three-dimensional welding. Proc. Inst. Mech. Eng. Part B J. Eng. Manuf. **212**(3), 175–182 (1998)
14. Clark, D., Bache, M.R., Whittaker, M.T.: Shaped metal deposition of a nickel alloy for aero engine applications. J. Mater. Process. Technol. **203**, 439–448 (2008)
15. Baufeld, B., Van der Biest, O., Gault, R.: Additive manufacturing of Ti–6Al–4V components by shaped metal deposition: microstructure and mechanical properties. Mater. Des. **31**(1), 106-S111 (2010). https://doi.org/10.1016/j.matdes.2009.11.032
16. Ibrahim, I.A., Mohamat, S.A., Amir, A., Ghalib, A.: The effect of gas metal arc welding (GMAW) processes on different welding parameters. Procedia Eng. **41**, 1502–1506 (2012). https://doi.org/10.1016/j.proeng.2012.07.342
17. Cao, Y., Zhu, S., Liang, X., Wang, W.: Overlapping model of beads and curve fitting of bead section for rapid manufacturing by robotic MAG welding process. Rob. Comput. Integr. Manuf. **27**, 641–645 (2011)
18. Leach, R., Thompson, A., Senin, N.: A metrology horror story: the additive surface. In: Proceedings of ASPEN/ASPE 2017 Spring Topical Meeting on Manufacture and Metrology of Structured and Freeform Surfaces for Functional Applications, 14–17 Hong Kong, China (2017)
19. Pagani, L., et al.: Towards a new definition of areal surface texture parameters on freeform surface: re-entrant features and functional parameters. Measurement **141**, 442–459 (2019). https://doi.org/10.1016/j.measurement.2019.04.027
20. Guerra, M.G., Gregersen, S.S., Frisvad, J.R., De Chiffre, L., Lavecchia, F., Galantucci, L.M.: Measurement of polymers with 3D optical scanners: evaluation of the subsurface scattering effect through five miniature step gauges. Meas. Sci. Technol. **31**, 1 (2019). https://doi.org/10.1088/1361-6501/ab3edb
21. Wieczorowski, M., Gapiński, B., Grzelka, M., Szostak, M., Szymański, M.: The use of photogrammetry in improving quality of workpieces after an injection molding process. Polymers **63**(2), 134–144 (2018). https://doi.org/10.14314/polimery.2018.2.7
22. Rękas, A., et al.: Analysis of tool geometry for the stamping process of large-size car body components using a 3D optical measurement system. Materials **14**, 7608 (2021). https://doi.org/10.3390/ma14247608
23. Affatato, S., Ruggiero, A., Logozzo, S.: Metal transfer evaluation on ceramic biocomponents: a protocol based on 3D scanners. Measurement **173**, 108574 (2021). https://doi.org/10.1016/j.measurement.2020.108574
24. Wieczorowski, M., Gapiński, B., Swojak, N.: The use of optical scanner for analysis of surface defects. In: Annals of DAAAM and Proceedings of the International DAAAM Symposium, vol. 30, no. 1, pp. 76–85 (2019). https://doi.org/10.2507/30th.daaam.proceedings.010

25. Majchrowski, R., Grzelka, M., Wieczorowski, M., Sadowski, Ł, Gapiński, B.: Large area concrete surface topography measurements using optical 3D scanner. Metrol. Meas. Syst. **22**(4), 565–576 (2015). https://doi.org/10.1515/mms-2015-0046
26. Yazdanbakhsh, S.A., Mohaghegh, K., Tiedje, N.S., De Chiffre, L.: Traceability of optical 3D scanner measurements on sand mould in the production of quality castings. Meas. Sci. Technol. **32**, 8 (2021). https://doi.org/10.1088/1361-6501/abf707
27. Gapiński, B., et al.: Use of white light and laser 3D scanners for measurement of mesoscale surface asperities. In: Diering, M., Wieczorowski, M., Brown, C.A. (eds.) MANUFACTURING 2019. LNME, pp. 239–256. Springer, Cham (2019). https://doi.org/10.1007/978-3-030-18682-1_19
28. Herráez, J., Martínez, J.C., Coll, E., Martín, M.T., Rodríguez, J.: 3D modeling by means of videogrammetry and laser scanners for reverse engineering. Measurement **87**, 216–227 (2016). https://doi.org/10.1016/j.measurement.2016.03.005
29. Lin, W., Shen, H., Fu, J., Wu, S.: Online quality monitoring in material extrusion additive manufacturing processes based on laser scanning technology. Precis. Eng. **60**, 76–84 (2019). https://doi.org/10.1016/j.precisioneng.2019.06.004
30. Swojak, N., Wieczorowski, M., Jakubowicz, M.: Assessment of selected metrological properties of laser triangulation sensors. Measurement **176**, 109190 (2021). https://doi.org/10.1016/j.measurement.2021.109190
31. Isa, M.A., Lazoglu, I.: Design and analysis of a 3D laser scanner. Measurement **111**, 122–133 (2017). https://doi.org/10.1016/j.measurement.2017.07.028
32. Blais, F., Beraldin, J.A.: Recent developments in 3D multi-modal laser imaging applied to cultural heritage. Mach. Vis. Appl. **17**, 395–409 (2006). https://doi.org/10.1007/s00138-006-0025-3
33. Hongyao, S., Jiaao, J., Bing, L., Zeyu, Z.: Measurement and evaluation of laser-scanned 3D profiles in wire arc hybrid manufacturing processes. Measurement **176**, 109089 (2021). https://doi.org/10.1016/j.measurement.2021.109089
34. Pavlenko, I., et al.: Parameter identification of cutting forces in crankshaft grinding using artificial neural networks. Materials **13**, 5357 (2020). https://doi.org/10.3390/ma13235357
35. Wieczorowski, M., Swojak, N., Pawlus, P., Pereira, A.: The use of drones in modern length and angle metrology. In: Śniatała, P., Iyengar, S.S., Bendarma, A., Klósak, M. (eds.) Modern Technologies Enabling Safe and Secure UAV Operation in Urban Airspace, pp. 125–140 (2021). IOS Press. https://doi.org/10.3233/NICSP210013
36. Carmignato, S., Dewulf, W., Leach, R. (eds.): Industrial X-Ray Computed Tomography. Springer, Cham (2018). https://doi.org/10.1007/978-3-319-59573-3
37. Kruth, J.P., Bartscher, M., Carmignato, S., Schmitt, R., De Chiffre, L., Weckenmann, A.: Computed tomography for dimensional metrology. CIRP Ann. Manuf. Technol. **60**, 821–842 (2011). https://doi.org/10.1016/j.cirp.2011.05.006
38. Michael, G.: X-ray computed tomography. Phys. Educ. **36**, 442–451 (2001)
39. Gapiński, B., Wieczorowski, M., Grzelka, M., Alonso, P.A., Tomé, A.B.: The application of micro computed tomography to assess quality of parts manufactured by means of rapid prototyping. Polymers **62**(1), 53–59 (2017). https://doi.org/10.14314/polimery.2017.053
40. Gapiński, B., Wieczorowski, M., Swojak, N., Szymański, M.: Geometrical structure analysis of combustible and non-combustible foams by computed tomography. J. Phys. Conf. Ser. **1065**(14) (2018). Art. no. 142025. https://doi.org/10.1088/1742-6596/1065/14/142025
41. Townsend, A., Pagani, L., Scott, P., Blunt, L.: Areal surface texture data extraction from X-ray computed tomography reconstructions of metal additively manufactured parts. Precis. Eng. **48**, 254–264 (2017). https://doi.org/10.1016/j.precisioneng.2016.12.008
42. Romano, S., Abel, A., Gumpinger, J., Brandao, A.D., Beretta, S.: Quality control of AlSi10Mg produced by SLM: metallography versus CT scans for critical defect size assessment. Addit. Manuf. **28**, 394–405 (2019). https://doi.org/10.1016/j.addma.2019.05.017

43. Hiller, J., Maisl, M., Reindl, L.M.: Physical characterization and performance evaluation of an x-ray micro-computed tomography system for dimensional metrology applications. Meas. Sci. Technol. **23**, 8 (2012). https://doi.org/10.1088/0957-0233/23/8/085404
44. Gapiński, B., Wieczorowski, M., Marciniak-Podsadna, L., Dybala, B., Ziolkowski, G.: Comparison of different methods of measurement geometry using CMM. Opt. Scanner Comput. Tomogr. 3D Procedia Eng. **69**, 255–262 (2014). https://doi.org/10.1016/j.proeng.2014.02.230

Verification of Computed Tomograph for Dimensional Measurements

Bartosz Gapiński[1(✉)], Michał Wieczorowski[1], Patryk Mietliński[1], and Thomas G. Mathia[2]

[1] Poznan University of Technology, M. Skłodowskiej 5 sq., 60965 Poznan, Poland
{bartosz.gapinski,michal.wieczorowski,
patryk.mietlinski}@put.poznan.pl
[2] École Centrale de Lyon, 36 Avenue Guy de Collongue, 69134 Écully CEDEX, France
thomas.mathia@ec-lyon.fr

Abstract. Radiation, is one of the ways of energy transfer by emission of electromagnetic waves or particles from a radioactive source. X-ray images make it possible to observe the internal structure of virtually any object and the only limitation is the ability to overexpose it. Tomography is a technique that allows imaging on the basis of sections or cross-sections obtained by means of a wave penetrating the object. The devices used in X-ray computed tomography consist of four basic components: the X-ray tube, the detector, the manipulator, and shielding to protect against the negative effects of X-rays on the operator. Measurements of geometric quantities carried out with a CT scanner are classified as coordinate measurements, and they are subject to the same conditions as, for example, contact or optical measurements carried out with other coordinate systems. Since that, the methods of its verification are derived from the methods applied to other devices operating in this technique. In the paper a metrological approach to computed tomography from verification and reverification point of view was presented. Documents related to it were briefly described and a practical example of inspection performed with standards was presented.

Keywords: Computed tomography · Verification · Standard

1 Introduction

Since the dawn of time, man has been trying to learn more about himself and the world around him. This desire is a mechanism of continuous development causing progress in all areas of life. Practically in every such case an indispensable "tool" is metrology in its broadest sense, giving the possibility of reliable description of reality with the help of definitions and units that are specific to it. Each correctly conducted measurement allows for confirmation or negation of the validity of theses and classification of phenomena or assessed objects. For centuries, in the field of metrology of geometric quantities, measurements could be divided into two main groups: contact measurements - that is, measurements where physical contact with the test object takes place, and non-contact measurements, where this contact does not occur. The results obtained by different techniques are not always consistent with each other [1]. In non-contact measurements, the most widely represented group are optical measurements. Apart from many advantages, their certain disadvantage is the necessity to observe the measured object, which excludes, for example, the evaluation of internal inaccessible closed spaces. In such a situation, the only solution some years ago was the necessity to destroy the tested object and the measurement, based on specially prepared sections.

Radiation, is one of the ways of energy transfer by emission of electromagnetic waves or particles from a radioactive source. One of the special features is that there is no need for a material medium between the source of emission and the receiver [2]. Taking into account the above arguments it can be stated that radiation penetrating the matter and enabling observation of internal structure of objects described by Wilhelm Röntgen was a breakthrough discovery. X-ray images make it possible to observe the internal structure of virtually any object and the only limitation is the ability to overexpose it. Thus, electromagnetic radiation can be used in metrology not only in the visible range [3]. X-rays discovered by Röntgen are electromagnetic waves [4]. The wavelength is assumed to be in the range $10^{-12} \div 10^{-8}$ m and can be further divided into more penetrating hard radiation ($10^{-12} \div 10^{-10}$ m) and less penetrating soft radiation ($10^{-10} \div 10^{-8}$ m). However, the boundaries set by wavelength are not sharp and in some areas the wavelength of X-rays overlaps with gamma rays.

X-rays - in addition to medical applications have also become important in technical sciences. They allow not only to find pores and identify different materials, but also to measure dimensions. However, in order to reproduce the measure correctly in this respect, metrological verification of the tomographs is necessary. Therefore, the paper presents steps in this direction and examples of activities-based.

2 Computed Tomography

In general, tomography is a technique that allows imaging on the basis of sections or cross-sections obtained by means of a wave penetrating the object. The image resulting from the measurement is obtained as an effect of reconstruction resulting from mathematical calculations. Tomography is used in all areas of science and life, and the choice of wave source depends on the application and safety conditions.

One of the best known varieties of tomography is X-ray based computed tomography (CT) or more precisely known as x-ray CT [5]. Its first application was related to medicine, however, it is gradually paving its way for application in technology and is increasingly used not only for defect analysis but also for measurements of geometrical features [6]. Computed tomography is called X-ray tomography in medical applications [7, 8]. In contrast, technical tomographs are often called microtomographs (micro-CTs), as they allow imaging with a resolution even below a micrometer. Since they are not designed for the study of living organisms, it is possible to increase the exposure time to X-rays by increasing the number of images, increasing the recording time, and the power of the X-ray tube. Micro-CTs are also used in medical and biological measurements when images with adequate resolution are sought [9, 10].

In X-ray CT measurement, the radiation beam is attenuated as it passes through the object, and this process depends on the thickness of the absorbing medium and the absorption coefficient related to the same length units. During the measurement, a number (usually hundreds or thousands) of 2D X-ray images are taken for different angular positions of the lamp-detector system relative to the measured object [11]. As a result of reconstruction from 2D shots, a spatial image is obtained. X-ray tomography is classified as non-destructive testing. However, one should always remember about the negative influence of X-rays on living organisms.

Of the tomographic techniques, X-ray computed tomography is the most widely used worldwide [12]. Allan MacLeod Cormack and Godfrey Newbold Hounsfield are considered to be its creators. The devices used in the technique are characterized by better resolutions and/or higher powers of radiation sources [13]. They consist of four basic components: the X-ray tube, the detector, the manipulator, and shielding to protect against the negative effects of X-rays on the operator. In a more automated version, the tomograph can also be equipped with a robot that feeds and receives objects for measurement, as shown in Fig. 1.

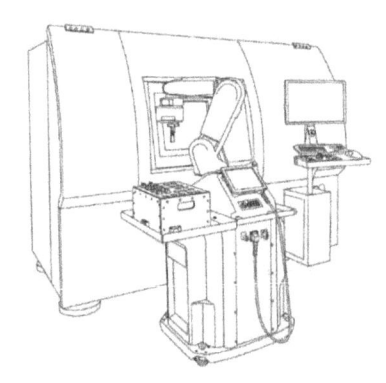

Fig. 1. Schematic image of the tomograph with the supply robot [14].

Since the beginning of CT as the youngest branch of coordinate measurement technology, attempts have been made to determine the parameters characterizing the performance of these devices from the point of view of accuracy features [15]. X-ray techniques previously used exclusively for the analysis of defects and flaws were limited to quality functions (good/bad) and the assessment of the object as in accordance with the requirements (flaws - if they could be detected, they are small enough not to disqualify the object) or not (defects occurred that are large enough to make the object unusable). This assessment was very often made by an operator, rarely supported by decision-making software, having a possibility of comparing a defect found in an object to a defect classified in a limit defects catalog. The development of tomographic techniques has allowed their use for both evaluation and measurement in a wide variety of fields, including metal components [16], plastics [17], foams [18], and even building components [19]. They have also become an important tool for verification of work in reverse engineering [20] and in the multi-criteria evaluation of components made by incremental techniques.

Measurements of geometric quantities carried out with a CT scanner are classified as coordinate measurements, and they are subject to the same conditions as, for example, contact or optical measurements carried out with other coordinate systems [21]. Similar conditions apply from ISO 17025 point of view, which concerns accredited testing and calibration laboratories.

3 Errors and Different Concepts of CT Verification

Since the technical tomograph is a device operating as a coordinate measurement system (CMS), in a natural way, the methods of its verification are derived from the methods applied to other devices operating in this technique. In this respect, the basic functionality of the tomograph covers the macro scale, although there are situations in which it can also be used for analyses on the micro scale [22, 23], as well as for data preparation for surface modelling [24]. The essential standard in this field describing methods for checking coordinate measuring systems in the macro scale is ISO 10360, whose individual parts concern, among others, verification of contact coordinate measuring machines [25], optical devices [26, 27], laser trackers [28], or articulated arms [29]. A part of standard for computed tomographs is currently in preparation and this part will have the number 11. But until it is published, the basic document describing the operation and metrological verification of a technical computed tomograph is the German recommendation VDI/VDE 2630.

The four parts of this elaboration [30–33] comprehensively discuss issues related to the design, capabilities, and accuracy characteristics of computer tomographs used for geometric analysis. This document was prepared by the German industry, which first saw in tomographs the opportunity to realize even more thorough dimensional inspection of objects and assembly groups, thanks to the active Association of German Engineers (VDI) and the Association of Electrical, Electronic and Computer Engineering (VDE). Its primary purpose is to define conditions and methods to ensure the comparability and traceability of measurement results performed with CT. This document was created in 2009–2010 and, as already mentioned, until now it is the only one addressing so comprehensively the issues presented above. Despite the passage of almost ten years, during which individual elements of CT scanners have developed, sometimes very dynamically, this study has not lost its relevance.

A number of factors influence the reliability of tomographic measurements. They can be divided according to their source, so we distinguish between factors related to the device itself (tomograph), the measurement task, the analysis procedure, as well as the environmental conditions and the operator. Among the factors related to the measurement device, we can further distinguish between factors originating from different elements of the tomograph. Those related to the X-ray tube, i.e. the radiation source, are primarily voltage and current and their stability over time, as well as the size and shape of the spot in focus, its position and stability, the material used for the material filters and their thickness, and any kind of abnormal beam propagation. The factors that cause inaccuracies related to lamp performance refer to drifts, namely, focus drift, lamp output drift, and temperature drift. Linear and rotational axis inaccuracies are another component of the device that affects its reliability. These include the orientation of the rotary axes, the perpendicularity of the source axis and the detector plane, interference from control and heating, and static and dynamic guidance errors of the linear and rotary axes.

The detector and the errors associated with it are also a very important element affecting the accuracy parameters of the CT scanner. There are a lot of different influencing factors, the most significant of which are internal scattered radiation, filter, cooling, pixel size and number of pixels (including errors), grey scale resolution, exposure time, stability, and even operating mode. When analyzing a CT scanner, one must also not forget about its environment and the changes that occur within it. Ambient conditions have such an impact as on any length measurement regardless of the measurement system used and refer not only to temperature and humidity, but also to vibrations and contamination in the air during the measurement and the influence of scattered radiation. On the side of the measurement task, the accuracy characteristics are of course influenced by the object itself and the measurement conditions, i.e. the parameters set on the tomograph. The object is characterized by both geometry (shape) and material data (ease of radiation penetration through the object), and influencing factors further include mounting, angular position, scattering and radiation hardening. On the other hand, pre-filtration, number of angular positions and vertical resolution, magnification and object position are among the important conditions. The analysis procedure, i.e., the capabilities and operation of the reconstruction software, as well as the data analysis related to the voxel size, are also crucial for the tomographic accuracy parameters and measurement uncertainty. Here we distinguish between voxel and surface data reduction, surface extraction, as well as object basing and the functioning of algorithms for reconstruction, correction and analysis. The last group of influencing factors are those resulting from the operator's work, i.e. the selection of the measurement strategy (resulting from the specific measurement task) and its implementation and experience, especially important in the case of items made of materials of very different densities and complex shapes. In addition, many of the factors described above influence not only the measurement uncertainty itself, but also other factors, resulting in complex relationships and the occurrence of dependent variables.

The standardization approach and the idea of creating a standard for the verification of tomographs as part of ISO 10360 includes - also from a standardization point of view - computer tomographs into the group of coordinate measuring systems and unifies the metrological approach to the verification of the functioning of these systems. The tomograph will thus be able to be a stand-alone measurement system, but it will also be able to be part of a multi-sensor system [34], and its errors will be governed by the relevant normative provisions. Treating the verification of these devices as part of ISO 10360 also makes it possible and necessary to use terms generally accepted throughout the standard. Part 11 of the ISO standard is therefore intended to give the definitions of the metrological characteristics and methods of their verification for coordinate measuring systems using the principle of computed tomography intended for the measurement of technical objects and appearing as a single-sensor device (if the system has two tubes it is treated like two lenses in optical coordinate measuring machines or two sets of lenses for different measuring fields for optical coordinate measuring scanners). The listing of technical subjects is intended to separate tomographs for industrial use from medical imaging and measuring devices and from tomographic material applications (e.g., defect analysis). Many standardization ideas related to standards are based on previous research work [35, 36]. The characteristics likely to be found in this part of the standard will therefore allow the specification of the parameters of coordinate measuring systems using the principle of computed tomography (attenuation of the signal when passing through material elements of different density) and any related comparisons. This applies to both planar and conical beam measuring systems, as well as systems collecting data along a spiral, which are considered by some to be the future of dimensional control, for example in the analysis of the wall thickness of castings. It is further assumed that the standards used for verification of CT-based coordinate measuring systems will be homogeneous (without obvious gradients in X-ray attenuation) and made of uniform materials. In addition, the effect of surface irregularities is determined to be negligible and not the subject of this part of the standard. However, it can also be applied to other CT systems, after appropriate adaptation and mutual acceptance (supplier/customer). Thus, the introduced standard addresses length and sampling errors. Additional aspects in the calibration of the systems described therein are expressed, among others, in the use of alternative length standards to gauge blocks and the comparability of features when using these standards, and the comparability of features when using different measurement strategies. It also introduces the term "artifact" as an error in an image being a consequence of the use of the term in the field of computed tomography. Errors that verify CT performance can be unidirectional and bidirectional. As elsewhere in the standard, any error should be less than the maximum allowable error. Maximum allowable values are generally provided by the coordinate measurement system manufacturer. This is always the case for new systems; if it is a reverification, the user can specify different values. Verification and reverification tests can also be carried out with a load on the load cell of the maximum mass allowed for a particular system. The manufacturer can also specify a maximum table load, expressed per unit area. In addition, the measurement time is important in the verification of CT systems and should be specified after verification. The manufacturer should specify a maximum time for the test and optionally also a minimum time.

4 Research Setup

A practical way of verifying the basic accuracy parameters of a CT scanner was realized by means of sample standards. A Waygate Technologies v|tome|x s 240 CT scanner equipped with two 240 kV/320 W microfocus and 180 kV/15 W nanofocus X-ray tubes was selected for the study. A temperature stabilized DXR 250RT detector array with 20 fps for real-time inspection, 200 μm pixel size, 1000 × 1000 pixels on a 200 mm × 200 mm large active area with 2x virtual detector enlargement was used to capture the X-ray images. It enables measurements of components made of both plastics and metals. As the CT scanner is classified as a coordinate measuring system, it was decided to carry out tests on a dedicated ball-bar type gauge with a sphere diameter of 5 mm and centre distance of 39.9715 ± 0.0010 mm. These two ruby spheres have the same nominal diameter. However, during the CT scan, one of them (sphere 1 at the top) is visible all the time, while sphere 2 (at the bottom) is sometimes obscured by the rod (pin). This situation may affect the obtained results. The second measurement cycle was to test gauge blocks with nominal sizes of 1, 5, 10 and 20 mm (Fig. 2). Gauge blocks are commonly used for dimensional inspection of CMMs, among others, so their use in this case is an analogy between CT and CMS.

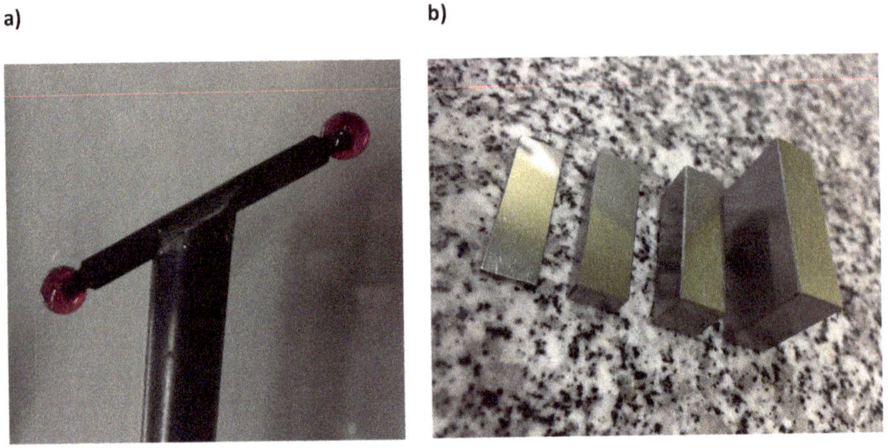

Fig. 2. Test standards used in the study: a) Ball-bar, b) 1, 5, 10 and 20 mm gauge blocks.

As the measured object moves away from the x-ray tube toward the detector, the magnification decreases. The ball-bar pattern was measured at three positions. In the first case, the standard was examined for the maximum possible position corresponding to the situation in which the object occupies the entire available detector space. This corresponds to a voxel size of 43.993 μm. The second position corresponded to the lowest possible magnification (for a feature distance of 500 mm from the lamp) and in this case the voxel has a size of 123.007 μm. The third position corresponded to the middle position between the two extremes, and the voxel size was 83.616 μm. In each case, measurements were made with a current of 190 μA, a voltage of 170 kV, and an exposure time of 200 ms for a single image. For each position, the measurement was performed with a different number of measurement images (1500, 1000, 800, 600, 400, 200, 100 images) uniformly distributed over a full rotation of the sample. The measurement for each condition was repeated 10 times.

In the second study, gauge blocks were evaluated. All elements were measured for the same X-ray tube parameters, i.e., 220 kV voltage and 245 μA current. The magnification was 5.099, which corresponds to a voxel size of 39.221 μm. For each gauge block, 1000 measurement shots were taken and each measurement was again repeated 10 times.

5 Results and Discussion

5.1 Ball-Bar Measurements

For each standard measured according to the procedure given, the diameter and sphericity of each of two spheres with a nominal diameter of 5 mm was determined. In addition, the distance between the centres of the reference spheres was calculated. Figure 3 shows the results for the diameter measurement of spheres 1 and 2.

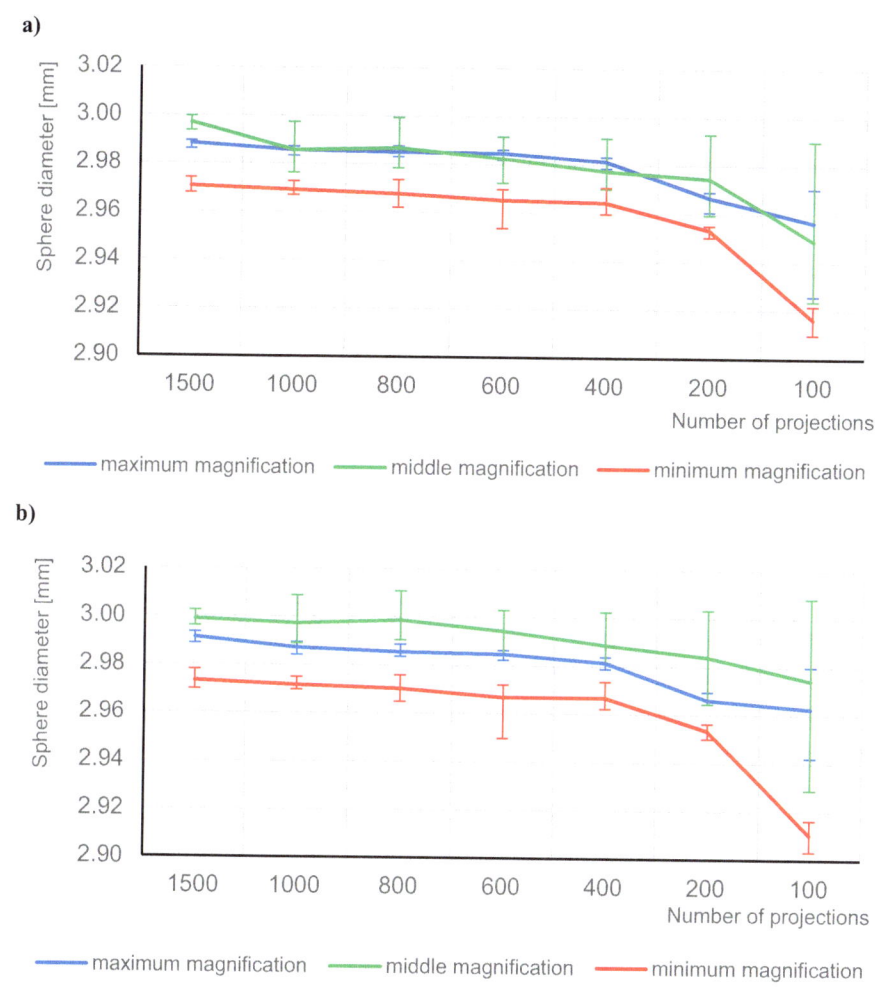

Fig. 3. Results of diameter measurement: a) sphere 1; b) sphere 2.

Analyzing the data presented in Fig. 3 it can be stated that in both cases the spheres measured with the lowest magnification are also characterized by the lowest diameter value. Also for those spheres the decrease of the diameter is most distinct, which is correlated with the reduction of the number of measurement images. The range of the diameter values for the individual number of shots is similar and reaches about 0.005 mm. The smallest values of the range can be observed for spheres measured with the maximum magnification, while the largest values were recorded for measurements with a small number of measurement images. It should be added here that 100 or even 200 shots are very small numbers and in practice may give far insufficient information about the measured object. In all cases the influence of the rod is not visible.

Figure 4 shows the shape deviation (sphericity) values for both measured reference spheres. In this case, some stability can be observed in the shape deviation results, for the standard measured with maximum and minimum magnification. This occurs in the range of 1500–600 measurement images. For the medium magnification there is a sharp increase in the shape deviation value between 1500 and 100 shots. Then its stabilization is observed up to 600 measurement images. In each of the measurement variants at the number of 400 images and less there is a sharp increase in the sphericity error resulting from less and less accurate reproduction of the shape of the reference spheres.

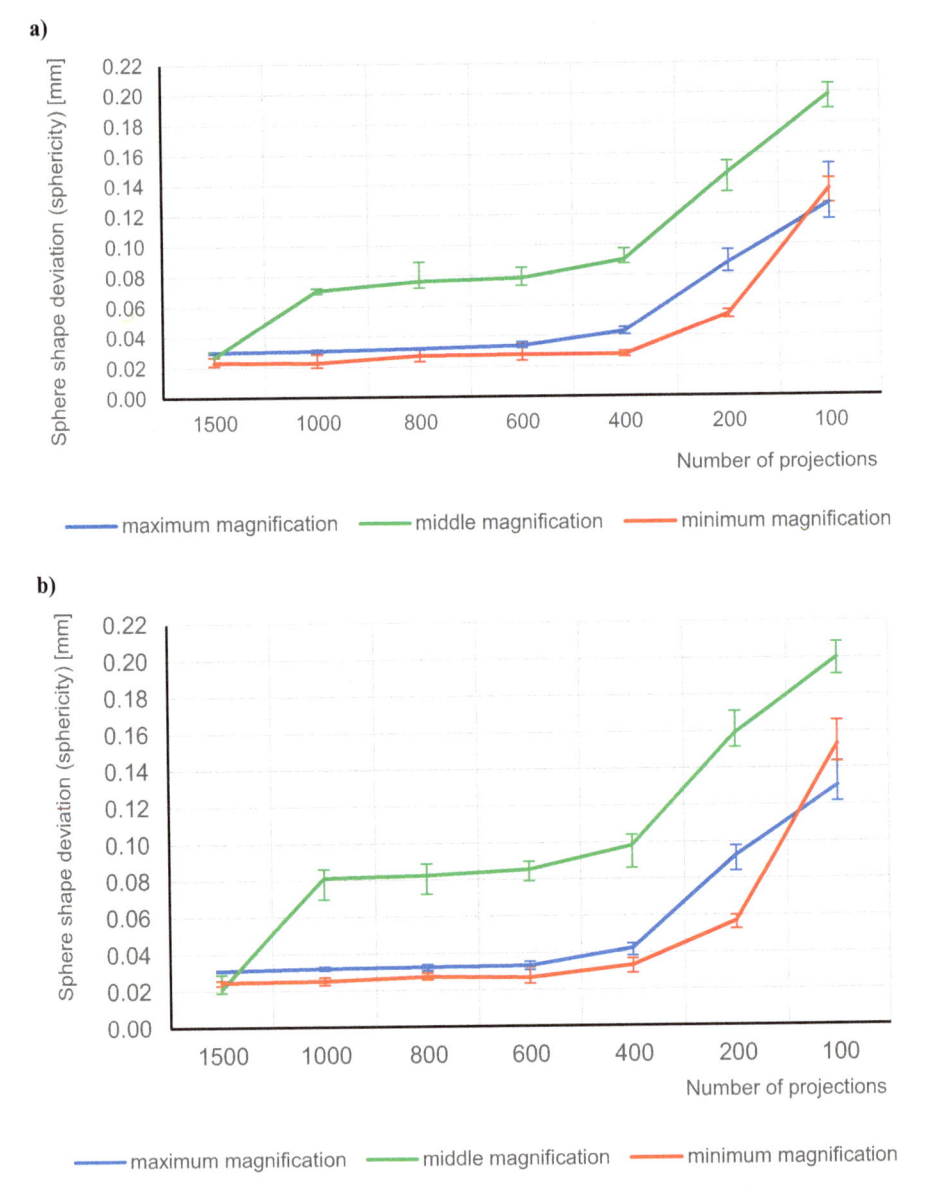

Fig. 4. Results of form measurement: a) sphere 1; b) sphere 2.

The last parameter that was determined for the ball-bar was the distance between the centres of its spheres. The results are shown in Fig. 5. Analyzing the data presented on the graph it can be observed that the difference of the average dimension in relation to the real value does not exceed ±0.005 mm in the range of 1500-400 measuring images. In the case of 200 and 100 images these values change significantly, which confirms that measurement with such a small number of projections can lead to erroneous information. When measuring the standard for the average magnification value, significant deviations in the obtained measurement results were observed. Sphericity is similar for both spheres - the measurement parameters were chosen correctly and the influence of the rod is not noticeable.

Tests conducted for a ball-bar type standard demonstrate its usefulness in the evaluation of CT scanner accuracy. What is important, such a construction of the standard allows to obtain a very wide range of standards in terms of dimensions, allowing verification of devices with different measurement ranges.

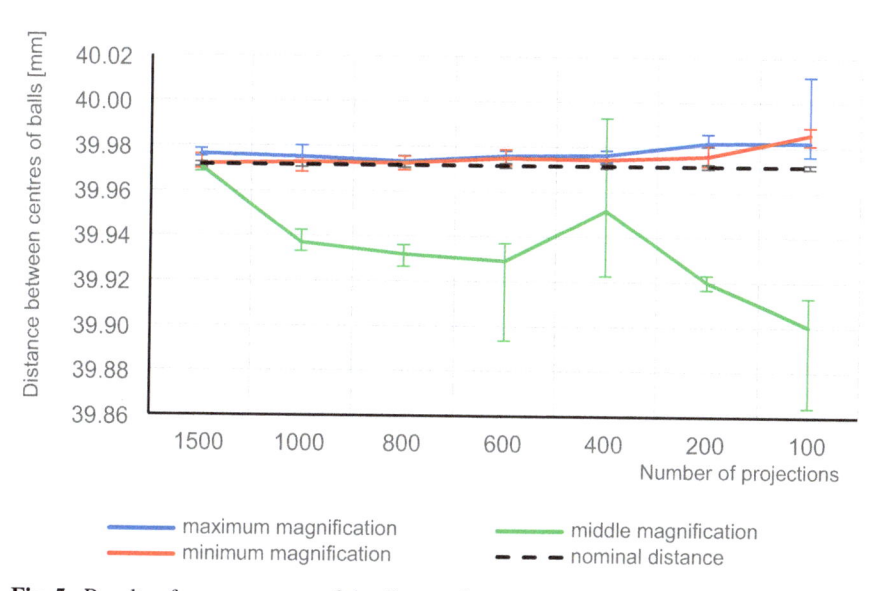

Fig. 5. Results of measurements of the distance between the centres of the ball-bar spheres.

5.2 Measurements of the Gauge Blocks

According to the methodology presented in chapter 4, gauge blocks with lengths of 1, 5, 10 and 20 mm were evaluated. The centre length of the gauge block was calculated, and the results are shown in Fig. 6.

Analyzing the obtained results, a strong increase in the dimensional difference can be observed as the length of the gauge block increases. This is due to the difficulty of penetration of "thick" gauge blocks by X-rays. During the passage through the measured object, there is a strong attenuation of the beam and an increase in the beam hardening

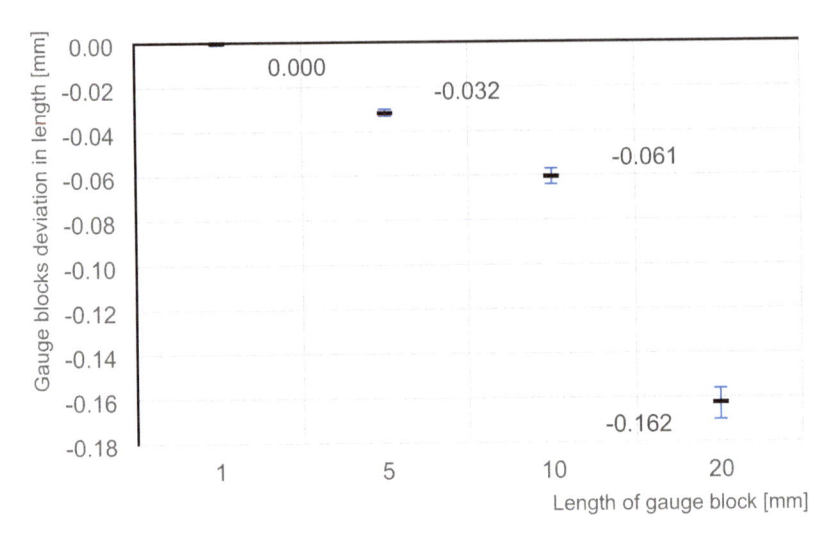

Fig. 6. Results of gauge blocks measurements.

phenomenon. This clearly shows that the use of gauge blocks to evaluate the accuracy of CT scanners is not proper and standards that do not cause such a significant obstruction during X-raying should be used.

6 Conclusions

From a verification and reverification point of view, it seems interesting to approach tomography as a single type of measurement and to be able to combine the results obtained by this technique with other coordinate methods, both optical and contact. In this way, it is possible to achieve smaller measurement uncertainty in areas where it is of particular interest, and to obtain a full set of coordinates (with larger uncertainty) in the rest of the measured object. This type of data fusion is possible not only at the macro scale, but also between scales, using macro-scale tomography and micro- or meso-scale surface roughness analysis techniques. Solutions combining traditional contact CMM and optical techniques and tomography in a single measuring device are also possible.

In this case the verification and reverification of the accurate functioning of the device requires the use of the guidelines of the already mentioned part 9 of ISO 10360. This standard distinguishes certain errors of a device or multisensor assembly, the systematics of which are similar to those used in its other parts. When analyzing the performance of multisensor assemblies, it has to be taken into account that not every software installed on a measuring system may have all the functions necessary to determine the geometric elements resulting from the definition of individual errors. In such a case, one can use a substitute evaluation or make an evaluation in external software.

Metrological approach to computed tomography allows fully reliable application of these devices for metrological analyses related to length and angle, and to some extent also to topography parameters in micro scale. Computed tomography also allows for the evaluation of surface irregularities in the way which is unattainable for other devices, as it makes it possible to assess reentrant features on the surface of printed elements made of sintered metal powders [37]. The influence of temperature on the measurement results is also very important with tomography, as in other parts related to coordinate techniques on a macro and micro scale [38]. This is especially important considering the multitude and variety of tasks that tomographs face. Looking into the future, the development of this measurement technology will further increase its application possibilities and it can be assumed with high probability that this technique should be expected to develop in the future similar to computed tomography in medicine.

Acknowledgments. This research was funded by the Polish Ministry of Higher Education grants No. 0614/SBAD/1547.

References

1. Pawlus, P., Reizer, R., Wieczorowski, M.: Comparison of results of surface texture measurement obtained with stylus methods and optical methods. Metrol. Meas. Syst. **25**(3), 589–602 (2018). https://doi.org/10.24425/123894
2. Banhart, J.: Advanced Tomographic Methods in Materials Research and Engineering. Oxford University Press (2008)
3. Bushong, S.: Computed Tomography. McGraw-Hill (2000)
4. Buzug, T.M.: Computed Tomography: From Photon Statistics to Modern Cone-Beam CT. Springer, Heidelberg (2008). https://doi.org/10.1007/978-3-540-39408-2
5. Kalender, W.A.: X-ray computed tomography. Phys. Med. Biol. **51**, 29–43 (2000). https://doi.org/10.1088/0031-9155/51/13/R03
6. Carmignato, S., Dewulf, W., Leach, R. (eds.): Industrial X-Ray Computed Tomography. Springer, Cham (2018). https://doi.org/10.1007/978-3-319-59573-3
7. Norman, D., Newton, T.H.: Localization with the EMI scanner. Am. J. Roentgenol. **125**(4), 961–964 (1975)
8. Paulus, M.J., Gleason, S.G., Kennel, S.J., Hunsicker, P.R., Johnson, D.K.: High resolution X-ray computed tomography: an emerging tool for small animal cancer research. Neoplasia **2**, 62–70 (2000)
9. Dąbrowski, M., Rogala, P., Uklejewski, R., Patalas, A., Winiecki, M., Gapiński, B.: Subchondral bone relative area and density in human osteoarthritic femoral heads assessed with micro-CT before and after mechanical embedding of the innovative multi-spiked connecting scaffold for resurfacing THA endoprostheses: a pilot study. J. Clin. Med. **10**(13), 2937–1–2937–17 (2021). https://doi.org/10.3390/jcm10132937
10. Kaczmarek, J., Bartkowiak, T., Paczos, P., Gapiński, B., Jąder, H., Unger, M.: How do the locking screws lock? A micro-CT study of 3.5-mm locking screw mechanism. Vet. Comp. Orthop. Traumatol. **33**(05) (2021). https://doi.org/10.1055/s-0040-1709728
11. Michael, G.: X-ray computed tomography. Phys. Educ. **36**, 442–451 (2001)
12. Elliott, J.C., Dover, S.D.: X-ray microtomography. J. Microsc. **126**(2) (1982). https://doi.org/10.1111/j.1365-2818.1982.tb00376.x
13. Sauerwein, C., Sim, M.: 25 years industrial computer tomography in Europe. In: Proceedings of CT-IP2003, DGZfP- BB 84-CD, Berlin, Germany (2003)
14. Wieczorowski, M., Swojak, N., Pawlus, P., Pereira, A.: The use of drones in modern length and angle metrology. In: Modern Technologies Enabling Safe and Secure UAV Operation in Urban Airspace, pp. 125–140 (2021). https://doi.org/10.3233/NICSP210013

15. Kruth, J.P., Bartscher, M., Carmignato, S., Schmitt, R., De Chiffre, L., Weckenmann, A.: Computed tomography for dimensional metrology. CIRP Ann. Manuf. Technol. **60**, 821–842 (2011). https://doi.org/10.1016/j.cirp.2011.05.006
16. Neuschaefer-Rube, U., et al.: Dimensional measurements with micro-CT-test procedures and applications. In: Proceedings of the 'Microparts' Interest Group Workshop, 28–29 October 2009. NPL, Teddington (2009). www.npl.co.uk/upload/pdf/091028_microparts_neuschaefer-rube.pdf
17. Gapiński, B., Wieczorowski, M., Grzelka, M., Alonso, P.A., Tomé, A.B.: The application of micro computed tomography to assess quality of parts manufactured by means of rapid prototyping. Polymers **62**(1), 53–59 (2017). https://doi.org/10.14314/polimery.2017.053
18. Gapiński, B., Wieczorowski, M., Swojak, N., Szymański, M.: Geometrical structure analysis of combustible and non-combustible foams by computed tomography. J. Phys: Conf. Ser. **1065**(14), 142025 (2018). https://doi.org/10.1088/1742-6596/1065/14/142025
19. Vanhove, Y., Wang, M., Wieczorowski, M., Mathia, T.G.: Metrological perspectives of tomography in civil engineering. In: 6th World Congress in Industrial Process Tomography, pp. 1389–1396 (2010)
20. Gapiński, B., Wieczorowski, M., Bak, A., Domínguez, A.P., Mathia, T.: The assessment of accuracy of inner shapes manufactured by FDM. In: AIP Conference Proceedings, vol. 1960, p. 140009 (2018). https://doi.org/10.1063/1.5035001
21. Gapiński, B., Wieczorowski, M., Marciniak-Podsadna, L., Dybala, B., Ziolkowski, G.: Comparison of different methods of measurement geometry using CMM. Optical Scanner and Computed Tomography 3D. Procedia Eng. **69**, 255–262 (2014). https://doi.org/10.1016/j.proeng.2014.02.230
22. Townsend, A., Pagani, L., Scott, P., Blunt, L.: Areal surface texture data extraction from X-ray computed tomography reconstructions of metal additively manufactured parts. Precis. Eng. **48**, 254–264 (2017). https://doi.org/10.1016/j.precisioneng.2016.12.008
23. Hamrol, A., Ciszak, O., Legutko, S., Jurczyk, M. (eds.): Advances in Manufacturing. LNME, Springer, Cham (2018). https://doi.org/10.1007/978-3-319-68619-6
24. Pawlus, P., Reizer, R., Wieczorowski, M.: A review of methods of random surface topography modelling. Tribol. Int. **152**, 106530 (2020). https://doi.org/10.1016/j.triboint.2020.106530
25. ISO 10360-2: Geometrical product specifications (GPS) - Acceptance and reverification tests for coordinate measuring systems (CMS) - Part 2: CMMs used for measuring linear dimensions (2009)
26. ISO 10360-7: Geometrical product specifications (GPS) - Acceptance and reverification tests for coordinate measuring systems (CMS) - Part 7: CMMs equipped with imaging probing systems (2011)
27. ISO 10360-8: Geometrical product specifications (GPS) - Acceptance and reverification tests for coordinate measuring systems (CMS) - Part 8: CMMs with optical distance sensors (2013)
28. ISO 10360-10: Geometrical product specifications (GPS) - Acceptance and reverification tests for coordinate measuring systems (CMS) - Part 10: Laser trackers for measuring point-to-point distances (2016)
29. ISO 10360-12: Geometrical product specifications (GPS) - Acceptance and reverification tests for coordinate measuring systems (CMS) - Part 12: Articulated arm coordinate measurement machines (CMM) (2016)
30. VDI/VDE 2630 Blatt 1.3: Computed tomography in dimensional measurement, Guideline for the application of DIN EN ISO 10360 for coordinate measuring machines with CT-sensors, Dusseldorf (2009)
31. VDI/VDE 2630 Part 1.1.: Computed tomography in dimensional measurement, Basics and definitions, Dusseldorf (2009)

32. VDI/VDE 2630 Part 1.2.: Computed tomography in dimensional measurement, Influencing variables on measurement results and recommendations for computed tomography dimensional measurements, Dusseldorf (2010)
33. VDI/VDE 2630 Part 1.4.: Computed tomography in dimensional measurement, Measurement procedure and comparability, Dusseldorf (2010)
34. ISO 10360-9: Geometrical product specifications (GPS) - Acceptance and reverification tests for coordinate measuring systems (CMS) - Part 9: CMMs with multiple probing systems (2013)
35. Müller, P.: Use of reference objects for correction of measuring errors in X-ray computed tomography. Kgs. Lyngby: DTU Mechanical Engineering, Technical University of Denmark, Copenhagen (2010)
36. Wang, D., Chen, X., Wang, F., Shi, Y., Kong, M., Zhao, J.: Measurement of size error of industrial CT system with calotte cube. In: Tan, J., Wen, X. (eds.) Proceedings of the 9th International Symposium on Precision Engineering Measurement and Instrumentation, Proceedings of SPIE, vol. 9446 (2015). https://doi.org/10.1117/12.2180600
37. Maszybrocka, J., Gapiński, B., Dworak, M., Skrabalak, G., Stwora, A.: Modelling, manufacturability and compression properties of the CpTi grade 2 cellular lattice with radial gradient TPMS architecture. Bull. Pol. Acad. Sci. Tech. Sci. **61**(4), 719–727 (2019). https://doi.org/10.24425/bpasts.2019.130181
38. Miller, T., Adamczak, S., Świderski, J., Wieczorowski, M., Łętocha, A., Gapiński, B.: Influence of temperature gradient on surface texture measurements with the use of profilometry. Bull. Pol. Acad. Sci. Tech. Sci. **65**(1), 53–61 (2017). https://doi.org/10.1515/bpasts-2017-0007

Identification of the Sensitivity of FDM Technology to Material Moisture with a Fast Test

Adam Hamrol, Maciej Cugier, and Filip Osiński

Faculty of Mechanical Engineering, Poznan University of Technology, Poznan, Poland
`filip.osinski@put.poznan.pl`

Abstract. The growing popularity of the Fused Deposition Modeling method in the industry and among individual users motivates to constantly research its potential. In particular, the influence of the moisture of the material on the result of the process has not yet been sufficiently described. The article presents a test that allows to determine this effect in an experiment requiring a relatively small number of repetitions. As a case study, a test for identifying the effects of material temperature, ply thickness, extrusion rate and material flow on dimensional accuracy, tensile strength and surface quality was presented. As a result of the conducted research, the possibility of using the method with a constant change of factors in the development of parameters of additive manufacturing was confirmed. In the tests performed, a significant increase in the dimensions of the produced samples was also noted with the filament moisture content at the level of 0.74%. For the given humidity, a significant deterioration of the visual condition of the surface was also noted. The moisture of material does not significantly affect the mechanical strength of the samples.

Keywords: Fused Deposition Modeling · FDM · Filament moisture · Design of experiments

1 Introduction

Materials used in Fused Deposition Modeling (FDM) are mostly polymers characterized by significant hygroscopicity. When stored, such a material absorbs moisture from its surroundings and water molecules are attached to polymer molecules, forming an intermolecular bond. Such a connection may cause microcracks that weaken polymer fibers and form empty spaces between layers. This phenomenon creates problems in polymer processing and in the quality of manufactured parts, for example in [1–5]:

- extruder jamming - water absorbed in the filament evaporates while melting in the heater block, which may result in material supply interruption
- deterioration of mechanical properties of product (material strength)
- large discrepancies in dimensions
- defects and high surface roughness

© The Author(s), under exclusive license to Springer Nature Switzerland AG 2022
M. Diering et al. (Eds.): MANUFACTURING 2022, LNME, pp. 156–165, 2022.
https://doi.org/10.1007/978-3-031-03925-6_14

Material used for FDM processing is delivered in the form of a wire and is commonly referred to as filament. One of the most common filaments is made of a polymer known as ABS. The recommended moisture of this material is 0.2–0.3% [6, 7]. To reduce moisture absorption, it is recommended to store ABS filament in special containers. If the moisture is too high, the filament should be dried. The recommended drying time and drying temperature depend on the type of material properties. For ABS, they are on average about 6 h and 80 °C [8, 9].

The effect of filament moisture on the properties of FDM products depends to a large extent on the printing conditions, primarily on the type of printing device [10]. Therefore, it is reasonable to develop an easy, cheap and fast test to determine how significant this effect is and how much it depends on FDM parameters, such as filament temperature, layer thickness, and filament speed. This information can be viewed as a kind of process diagnostics [11] and should be helpful for an operator when setting the process parameters of the stability of the filament moisture is uncertain.

2 The Test Methodology

The test presented in this paper is based on an experiment with a systematic change of process parameters (P) [12]. These parameters are referred to as "factors" in the terminology of design of experiments [13]. The advantage of this type of experiment is the small number of trials needed to obtain reliable data on the influence of a large number of factors (i.e. four or more) on the results of the process. Its weakness is the lack of information on the interaction between factors, which can be identified, for example, in a factorial experiment. However, for the purpose of the presented test, these interactions are irrelevant. Such a limitation means that the test can be used mainly in the case of basic materials (filaments) that are used in additive manufacturing with the FDM method. Thanks to the application of the presented test, the operator can make an unambiguous decision regarding the modification of the process parameters or putting the filament for drying in order to obtain the required print parameters.

For each factor, two levels of its setting are determined:

- level (+) - giving potentially better process results (Y_B)
- level (−) - giving potentially worse process results (Y_W)

In the first part of the experiment, two trials are carried out:

1. with all factors on the level " + " (All (+))
2. with all factors on the level "−" (All (−))

Two statistics are calculated:

- the difference of average values obtained in the trials:

$$D = \overline{Y_B} - \overline{Y_W}$$

- the average range of results in each of the trials:

$$\overline{R} = \frac{R_B + R_W}{2}$$

If the ratio:

$$q = \frac{D}{\overline{R}} > q_c$$

it can be concluded that the difference (D) of averages $\overline{Y_W}$ and $\overline{Y_B}$ is a result of the process parameter settings. The value of q_c is adopted as the critical value of a test for averages difference in t-Student distribution.

Based on the value of $\overline{R}$ and for the given significance level α, upper and lower limits of confidence interval (UCL and LCL) of $\overline{Y_W}$ and $\overline{Y_B}$ are calculated.

In the second part of the experiment, two series of trials are carried out:

1. with all factors set on (+) but with one factor P_i set on (−)
2. with all factors set on (−) but with one factor P_i set on (+)

If the value Y in a trial is located outside the confidence interval of $\overline{Y_W}$ and $\overline{Y_B}$, it can be concluded that the influence of this factor on process outcomes is important.

The number of trials necessary to carry out the experiment is $n = 2k + 2$, where k is the number of process parameters investigated. Each trial should be replicated at least 3 times. For example, this means that $k = 3$, the number of trials is 8, and the number of all replications 18.

The test is carried out in the following steps:

1. Preparation of at least two sets of filaments – one of the moisture recommended by the manufacturer, and the other with a maximum moisture expected in the conditions in which it is stored.
2. Selection of properties important for the manufactured product.
3. Selection of process parameters that have significant impact on the product properties selected.
4. Scheduling the plan of the experiment.
5. Conducting a set of the planned experiment trials.
6. Calculation of the average values confidence intervals.
7. Analysis of the obtained results.

3 Case Study

3.1 Tested Object

The tests were carried out on a Zortrax M200 Plus additive manufacturing 3D printer. Samples of a standardized shape intended for mechanical tests PN-EN ISO 527-2:2012 were printed [8] (Fig. 1).

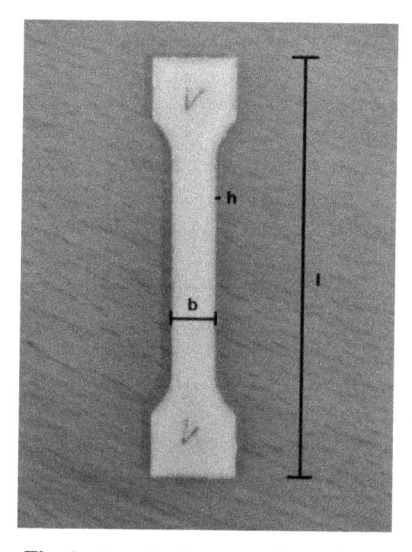

Fig. 1. Sample shape and dimensions.

The model of the sample for printing was designed in Autodesk Inventor, exported to .zprojx format, and launched in Z-SUIT, which allowed for the preparation of executive files for the Zortrax device.

The moisture content of the material for printing was tested on a RADWAG MA 50/1.R moisture analyzer.

3.2 Product and Process Parameters

The following properties of the sample were measured:

- sample width - B [mm].
- R_m tensile strength [MPa]
- appearance of the surface (visual evaluation); three states of surface quality were distinguished:

  1. good
  2. acceptable
  3. nonacceptable

The sample width (B) was measured by electronic caliper Mitutoyo 500-181-30, with a measuring accuracy to ± 0.01.

A statistical stretch test was carried out on a SunPoc WDW-5D-HS device according to the procedure described in EN ISO-527 [8, 9].

A Delta Optical Smart 5MP Pro microscope was used to assess the quality of the surface structure.

Four process parameters were adopted as experiment factors:

- *P1* - material temperature during extrusion
- *P2* - the thickness of the applied material layer
- *P3* - extrusion speed - nozzle moving speed, that extrudes the plastic relative to the table
- *P4* - material flow - amount of molten polymer flowing through the nozzle in the unit

Based on the results of previous tests [1] and literature data, the levels (+) and (−) of factors (i.e. process parameters) were defined (Table 1) at which the sample proper-ties should presumably be better (Y_B) or worse (Y_W).

Table 1. Parameter (i.e. factors) and levels of their setting.

Mark	Parameter type	(+)	(−)
P1	Material temperature	245 °C	255 °C
P2	Layer thickness	0.09 mm	0.39 mm
P3	Extrusion speed	27 mm/s (0%)	30 mm/s (10%)
P4	Filament flow	0%	10%

The ABS filament was tested in three humidity states: 0.25%, 0.34% and 0.75%.

3.3 Plan of the Experiment

The plan of the experiment is presented in Table 2.

Table 2. Plan of the experiment.t

Factor	No. of trial	Level	No. of trial	Level
All factors (+)	1a	All (+)	1b	All (−)
Hotend temperature	2a	P1 (−) All (+)	2b	P1 (+) All (−)
Layer thickness	3a	P2 (−) All (+)	3b	P2 (+) All (−)
Extrusion speed	4a	P3 (−) All (+)	4b	P3 (+) All (−)
Filament speed	5a	P4 (−) All (+)	5b	P4 (+) All (−)

The experiment was carried out according to the plan in Table 2. Three replications were performed in each trial.

3.4 Presentation of Results and Calculations

The results of the experiment in regard to the width, strength and surface quality of the sample are summarized in Table 3.

Identification of the Sensitivity of FDM Technology to Material Moisture

Table 3. Results of the experiment.

No. of trial	Factor level		Width - B [mm]			Strength R_m [MPa]			Surface quality 1, 2 or 3		
			Filament moisture								
			0.25%	0.32%	0.74%	0.25%	0.32%	0.74%	0.25%	0.32%	0.74%
1	a	All (+)	10.07	9.96	10.07	32.91	28.04	30.26	1	1	2
	b	All (−)	10.27	10.31	10.51	24.19	23.89	21.08	2	2	3
2	a	P1 (−) All (+)	10.06	10.04	10.12	31.6	32.29	29.22	1	2	2
	b	P1 (+) All (−)	10.27	10.28	10.38	23.83	23.97	21.57	1	2	2
3	a	P2 (−) All (+)	10.23	10.17	10.26	17.38	17.22	17.98	1	2	3
	b	P2 (+) All (−)	10.23	10.26	10.29	32.73	33.55	31.71	1	2	2
4	a	P3 (−) All (+)	10.01	10.02	10.11	30.24	28.12	30.36	1	2	3
	b	P3 (+) All (−)	10.26	10.28	10.44	24.43	22.33	20.67	1	1	2
5	A	P4 (−) All (+)	10.17	10.26	10.25	28.73	30.52	32.61	2	3	3
	b	P4 (+) All (−)	10.16	10.19	10.34	16.96	18.00	18.65	1	2	2

The results of the significance tests of the ratio q and limits of confidence intervals regarding the average values of width and strength of the sample are presented in Tables 4 and 5.

Table 4. Significance tests and confidence intervals for width

Moisture		Y	R	$\bar{R}$	D	q	$q_{kr} = 0.89$	UCL	LCL
							$\alpha = 0.05$; Student's t-distribution		
0.25%	All (+)	10.07	0.08	0.080	0.2	2.5	Significant	10.20	9.94
	All (−)	10.27	0.08					10.40	10.14
0.34%	All (+)	9.96	0.07	0.075	0.35	4.67	Significant	10.08	9.84
	All (−)	10.31	0.08					10.43	10.19
0.74%	All (+)	10.07	0.06	0.055	0.44	7.33	Significant	10.16	9.98
	All (−)	10.51	0.05					10.60	10.42

Table 5. Significance tests and confidence intervals for strength

Moisture		Y	R	$\bar{R}$	D	q	$q_{kr} = 0.89$	UCL	LCL
							$\alpha = 0.05$; Student's t-distribution		
0.25%	All (+)	32.91	0.61	0.480	8.72	18.16	Significant	33.67	32.15
	All (−)	24.19	0.35					24.95	23.43
0.34%	All (+)	28.04	1.28	1.480	4.15	3.29	Significant	30.40	25.68
	All (−)	23.89	1.68					26.25	21.53
0.74%	All (+)	30.27	3.0	2.350	9.18	3.91	Significant	34.01	26.53
	All (−)	21.09	1.7					24.83	17.35

In order to facilitate the analysis of the results obtained they are presented graphically in Figs. 2 (a–d), 3 (a–d) and 4.

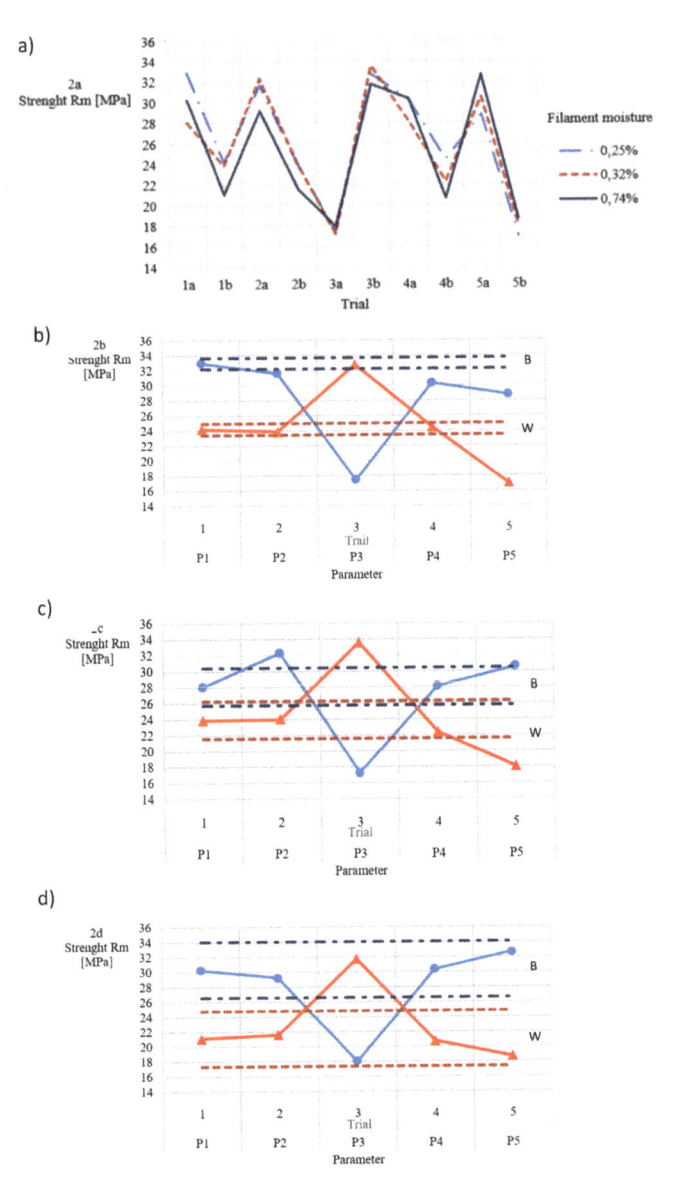

Fig. 2. Sample strength depending on the level of factors for different filament humidity

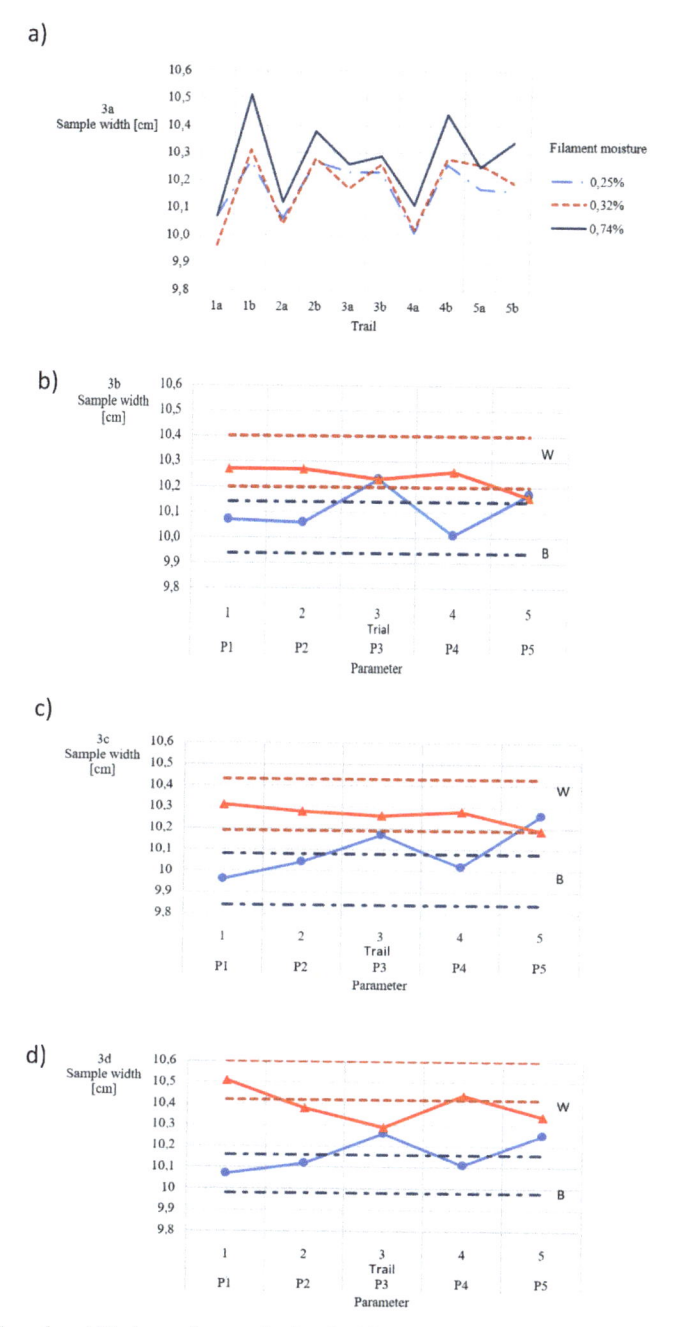

Identification of the Sensitivity of FDM Technology to Material Moisture

Fig. 3. Sample width depending on the level of factors for different filament humidity.

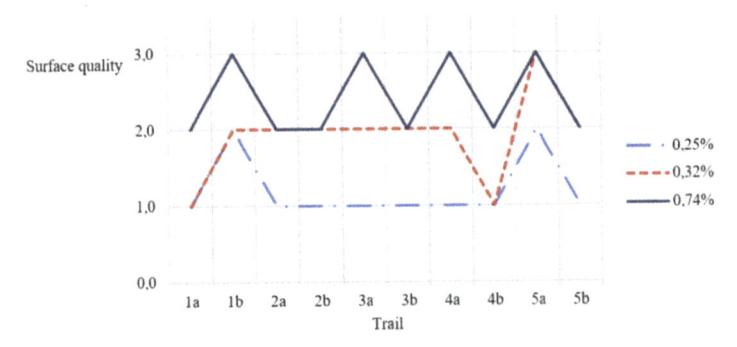

Fig. 4. Surface quality depending on the level of factors for different filament humidity.

4 Analysis of the Results

The results of the experiment can be used as a basis to decide about the need to dry filament, or to give guidance on parameter settings that allow you to reduce the negative impact of filament moisture on the printing results.

Material moisture had a significant impact on the dimensional accuracy of the samples. With an increase in filament moisture, the width of a sample increases, thus the difference between the actual dimension and the nominal dimension increases. However, the differences at moisture levels of 0.25% and 0.34% are quite irrelevant. A strong increase in the sample width for the 0.74% moisture level was measured. The dimensional accuracy of the sample affects the thickness of the layer and the speed of the filament supply.

The moisture of material does not significantly affect the mechanical strength of the samples. However, as with the width of a sample, the sample strength is affected by the layer thickness and filament flow speed – the highest strength is achieved at the parameters setting All (+) and P2 (+) All (−).

A deterioration of surface quality for samples printed with filament with 0.74% moisture is noticeable. The individual material paths and surface damage are more visible, which can be directly translated into a greater surface roughness.

The final decision on the need to dry filament depends on the process operator or the customer that needs the manufactured products. In this case study, it would be reasonable to dry the filament in cases where the dimensional accuracy and the surface quality of the product are crucial.

In addition, in a case of suspected high humidity of filament, and in the absence of proper preparation of material for production (i.e. drying), it is recommended to adjust the process using the material speed. Increasing this parameter, while maintaining a constant flow, causes in practice a reduction of the amount of material. It can be presumed that a similar effect can be obtained by decreasing the flow of filament through the nozzle, but in the above example, such a case was not analyzed.

References

1. Wichniarek, R., Hamrol, A., Kuczko, W., Górski, F., Rogalewicz, M.: ABS filament moisture compensation possibilities in the FDM process. CIRP J. Manuf. Sci. Technol. **35**, 550–555 (2021)
2. https://vshaper.com/pl/blog-pl/jak-wilgoc-wplywa-na-wlokno-drukarskie-3d/. Accessed 04 Aug 2021
3. Algarni, M.: The influence of raster angle and moisture content on the mechanical proper-ties of PLA parts produced by fused deposition modeling. Polymers **13**, 237 (2021). https://doi.org/10.3390/polym13020237(2020)
4. Cueto, C.I.A.: Anisotropy and humidity effect on tensile properties and electrical volume resistivity of fused deposition modeled acrylonitrile butadiene styrene composites. PUCP (2017)
5. https://blog.gotopac.com/2018/03/01/how-3d-printer-filament-storage-cabinets-instantly-improve-3d-print-part-quality/#Whats_Wrong_with_Drying_3D_Printer_Filament_by_Baking_It. Accessed 04 Aug 2021
6. https://filament2print.com/gb/blog/45_humidity-problems-3d-filaments.html. Accessed 04 Aug 2021
7. Broniewski, T., Kapko, J., Płaczek, W., Thomalla, J.: Test methods and evaluation of the properties of plastics. Wydawnictwo Naukowo-Techniczne, Warszawa (2000)
8. Standard Syntetics: Determination of mechanical properties at static stretching. PN-EN ISO 527-2:2012
9. Standard, Determination of mechanical properties at static stretching. PN-EN ISO 527-1:2012
10. Galeja, M., Hejna, A., Kosmela, P., Kulawik, A.: Static and dynamic mechanical properties of 3D printed ABS as a function of raster angle. PMCID: PMC7013835 (2020)
11. Hamrol, A.: Process diagnostic as a means of improving the efficiency of quality control. Prod. Plann. Control. **11**, 8 (2000)
12. Bhote, K.R.: Qualitaet – der Weg zur Weltspitze. IQM – Institut fuer Qualitaetsmanagement, Grossbottwar (1990)
13. Montgomery, D.C., Runger, G.C.: Applied Statistics and Probability for Engineers, 6th edn. Wiley, New York (2014)

Author Index

A
A. Rani, Ahmad Majdi, 10
A. Rani, M. Nasir, 10
Alejandro, Pereira Domínguez, 127
Awang, Mohd Norhisyam, 10

B
Budzik, Grzegorz, 127

C
Cepova, Lenka, 48, 79
Česánek, Zdeněk, 67
Cugier, Maciej, 156

D
Diering, Magdalena, 127

G
Gapiński, Bartosz, 89, 127, 142
Garashchenko, Yaroslav, 36
Gawlik, Józef, 1
Grabowski, Marcin, 1, 26

H
Hajnys, Jiri, 79
Hamrol, Adam, 156
Harmatys, Wiktor, 115
Houdková, Šárka, 67

I
Ishar, Nur Athirah, 10

J
Jakubowicz, Michal, 48

K
Kogan, Ilja, 36
Konieczna, Zuzanna, 57
Kowalczyk, Małgorzata, 1
Krajewska-Śpiewak, Joanna, 1
Krawczyk, Bartłomiej, 89
Kupiec, Robert, 115

M
Mathia, Thomas G., 142
Mazur, Tomasz, 48
Meijer, Frans, 57
Mesicek, Jakub, 79
Mietliński, Patryk, 142
Mizera, Ondrej, 79
Mohamad Zakir, Bushra, 10
Mohd Nuri Al-Amin, Nu'man Al-Basyir, 10

O
Osiński, Filip, 156
Ostrowska, Ksenia, 115

P
Polach, Pavel, 67

R
Rucki, Miroslaw, 36, 48
Ryba, Tomasz, 36
Rybicki, Marek, 102

S
Sanetra, Izabela, 115
Składanowska, Katarzyna, 115
Skoczypiec, Sebastian, 26

Smak, Krzysztof, 89
Stachowska, Ewa, 57
Subrmaniam, Krishnan, 10
Szablewski, Piotr, 89
Szostak, Marek, 10

T
Thompson, Harvey M., 10
Tyczyński, Piotr, 1

W
Wieczorowski, Michał, 127, 142
Wyszynski, Dominik, 26

Y
Yago, I. P., 127

Z
Zelinka, Jan, 79

CPSIA information can be obtained
at www.ICGtesting.com
Printed in the USA
LVHW081146210422
716831LV00001B/13

9 783031 039249